Wavelet-Based Vibration Control of Smart Buildings and Bridges

Wavelet-Based Vibration Control of Smart Buildings and Bridges

Hojjat Adeli

Hongjin Kim

CRC Press
Taylor & Francis Group
Boca Raton   London   New York

CRC Press is an imprint of the
Taylor & Francis Group, an Informa business

CRC Press
Taylor & Francis Group
6000 Broken Sound Parkway NW, Suite 300
Boca Raton, FL 33487-2742

© 2009 by Taylor & Francis Group, LLC
CRC Press is an imprint of Taylor & Francis Group, an Informa business

---

Library of Congress Cataloging-in-Publication Data

Adeli, Hojjat, 1950-
    Wavelet-based vibration control of smart buildings and bridges / Hojjat Adeli and Hongjin Kim.
        p. cm.
    "A CRC title."
    Includes bibliographical references and index.
    ISBN 978-1-4200-8923-3 (hardcover : alk. paper)
    1. Intelligent buildings--Vibration--Mathematical models. 2.
Bridges--Vibration--Mathematical models. 3. Active noise and vibration control.
4. Wavelets (Mathematics) I. Kim, Hongjin, 1969- II. Title.

TA654.A34 2009
624.1'762--dc22   2009010066

Visit the Taylor & Francis Web site at
http://www.taylorandfrancis.com

and the CRC Press Web site at
http://www.crcpress.com

Dedicated to

Nahid, Anahita, Amir, Mona, and Cyrus Dean
Adeli

Ellen and Jayoung
Kim

CONTENTS

PREFACE

Over the past fifteen years or so, the senior author has been advocating and advancing the vision and concept of *smart structure* where intelligent sensors and actuators are integrated for health monitoring and vibration control of structures under extreme external dynamic loading. The vision was launched with the publication of *Control, Optimization, and Smart Structures: High-Performance Bridges and Buildings of the Future* (with A. Saleh, John Wiley & Sons, 1999), and culminated with the recent publication of *Intelligent Infrastructure: Neural Networks, Wavelets, and Chaos Theory for Intelligent Transportation Systems and Smart Structures* (with X. Jiang, CRC Press, Taylor & Francis 2009). This book further advances the same vision.

In the conventional method of designing a structure such as a highrise building structure additional materials are used in the form of larger member sizes, bracings, and shear walls to resist the destructive forces of nature due to earthquakes or winds. The additional materials needed to resist seismic forces increase exponentially with the height of the structure. In the case of a superhighrise building structure, say a 120-story-tall building, the additional structural materials can be in the same order of magnitude needed to carry the vertical dead and live loads. In this conventional aseismic design approach the buildings are *sitting ducks*; they take the punishment from the earthquake forces with no *fight*.

In a *smart structure* sensors are placed strategically in the structure to measure the response of the structure and properly designed actuators are used to apply internal forces to compensate for the destructive forces of the nature. Such a smart structure will be substantially lighter than the corresponding conventional structure resulting in a more sustainable design. The key to the realization of such a smart structure technology is development of an effective control algorithm for determining the required magnitudes of control forces in real time accurately and reliably. In an attempt to develop a powerful algorithm created specifically for vibration control of complex and large civil structures the authors introduced the concept of wavelets in the field of structural control for the first time. This book presents the authors' pioneering research and technology on vibration control of civil structures.

The book also presents a semi-active tuned liquid column damper (TLCD) system. The original TLCD system, a relatively new idea, is a passive system where the size of the orifice is fixed. This system has recently been installed in a few highrise buildings, most recently in the 57-story Comcast Office Building in Philadelphia which opened in June 2008. In the semi-active TLCD system advanced in this book the size of the orifice is changed in real time using battery power for any external dynamic excitation. The new wavelet-based control algorithm is used to determine the size of the orifice in real time.

The validity of the new technology is demonstrated by application to realistic and large structures. The example applications include both building and bridge structures. It is well-known that irregular buildings are punished and often damaged significantly more than regular buildings during strong ground motions due to the complicated behavior of such structures. Consequently, the vibration control of such structures is particularly challenging. Until recently structural designers tended to choose regular forms for highrise buildings as observed, for example, in the Los Angeles skyline. But now clients demand unique landmark structures that are often irregular with complicated dynamic behavior, as seen in recent and forthcoming designs in various cities from Milan to Dubai. The efficacy of the new model is demonstrated through vibration response control of three dimensional irregular buildings under various seismic excitations.

The effectiveness of both a semi-active TLCD system and a hybrid damper-TLCD control system is also demonstrated for the control of wind-induced motion of a 76-story highrise building. Finally, the efficacy of the new wavelet-based control algorithm is shown for vibration control of a cable-stayed bridge under various seismic excitations.

ACKNOWLEDGMENT

The work presented in this book was partially sponsored by the National Science Foundation through research grants to the senior author. Parts of the work were published by the authors in several research journals: *Journal of Structural Engineering*, *Journal of Bridge Engineering* (both published by American Society of Civil Engineers), and *International Journal for Numerical Methods in Engineering* and *Computer-Aided Civil and Infrastructure Engineering* (both published by Wiley-Blackwell). Chapter 4 is based on a journal article written by the senior author and his former research associate, Ziqin Zhou, and is reproduced by permission of the publisher of the journal.

ABOUT THE AUTHORS

Hojjat Adeli received his Ph.D. from Stanford University in 1976 at the age of 26 after graduating from the University of Tehran in 1973 ranking number one in the entire College of Engineering. He is currently Professor of Civil and Environmental Engineering and Geodetic Science and the holder of the Abba G. Lichtenstein Professorship at The Ohio State University. He has authored over 440 research and scientific publications in various fields of computer science, engineering, applied mathematics, and medicine. He is the Founder and Editor-in-Chief of the international research journals *Computer-Aided Civil and Infrastructure Engineering*, in publication since 1986, and *Integrated Computer-Aided Engineering,* in publication since 1993. He is also the Editor-in-Chief of the *International Journal of Neural Systems.* He is the quadruple winner of The Ohio State University College of Engineering Lumley Outstanding Research Award. In 1998 he received the Distinguished Scholar Award, The Ohio State University's highest research award *"in recognition of extraordinary accomplishment in research and scholarship".* In 2005, he was elected Honorary/Distinguished Member, American Society of Civil Engineers *"for wide-ranging, exceptional, and pioneering contributions to computing in civil engineering disciplines and extraordinary leadership in advancing the use of computing and information technologies in many engineering disciplines throughout the world."* In 2006, he received the ASCE Construction Management Award *"for development of ingenious computational and mathematical models in the areas of construction scheduling, resource scheduling, and cost estimation."* In 2007, he received

The Ohio State University College of Engineering Peter L. and Clara M. Scott Award for Excellence in Engineering Education *"for sustained, exceptional, and multi-faceted contributions to numerous fields including computer-aided engineering, knowledge engineering, computational intelligence, large-scale design optimization, and smart structures with worldwide impact,"* as well as the Charles E. MacQuigg Outstanding Teaching Award. In 2008 he was elected Fellow of the American Association for the Advancement of Science for *"distinguished contributions to computational infrastructure engineering and for worldwide leadership in computational science and engineering as a prolific author, keynote speaker, and editor-in-chief of journals."* He has presented Keynote Lectures at 71 conferences held in 39 different countries. He has been Chair or Honorary Chair of 20 and a member of organizing or program committees of 286 conferences held in 58 countries. He holds a U.S. patent in the area of large-scale optimization.

Hongjin Kim received his B.S. in Architectural Engineering from Seoul National University in 1993, and M.S. and Ph.D. degrees in Civil (Structural) Engineering from The Ohio State University in 1999 and 2002, respectively. He was a researcher at the Research Institute of Industrial Science & Technology (RIST), Kyungkido, Korea. He has been an assistant professor at Kyungpook National University, Daegu, Korea since 2007. He has authored 18 papers published in the leading civil engineering journals such as the ASCE *Journal of Structural Engineering*, ASCE *Journal of Bridge Engineering*, and the *International Journal for Numerical Methods in Engineering*. He is interested in structural control against wind and seismic loads, structural system design of tall buildings, vibration analysis, and structural health monitoring. Current research directions include structural health monitoring of tall buildings using wireless sensors, inverse force identification based on the structural responses, and mitigation of wind-induced motion of tall buildings using liquid dampers.

Chapter 1
Introduction

1. 1. Motivation and objectives

In the traditional method of designing a structure such as a highrise building structure additional materials are used in the form of larger member sizes, bracings, and shear walls to resist the destructive forces of nature due to earthquakes or winds. The additional materials needed to resist seismic forces increase exponentially with the height of the structure. In the case of a superhighrise building structure, say a 120-story-tall building, the additional structural materials can be in the same order of materials needed to carry the vertical dead and live loads. In this conventional aseismic design approach the buildings are *sitting ducks*; they take the punishment from the earthquake forces with no *fight*.

In a smart structure sensors are placed strategically in the structure to measure the response of the structure and properly designed actuators are used to apply internal forces to compensate for the destructive forces of nature. Such a smart structure will be substantially lighter than the corresponding conventional structure resulting in a more sustainable design. The key to the success of such a smart structure technology is development of an effective control algorithm for determining the required magnitudes of control forces in real time accurately and reliably.

A large number of papers have been published on active vibration control of structures during the past three decades. The great majority of these papers, however, use control algorithms developed primarily in the aerospace industry such as the Linear Quadratic Regulator (LQR) feedback control algorithm and the Linear Quadratic Gaussian (LQG) control algorithm. The primary objective of this book is to present a new control algorithm for robust control of smart civil structures subjected to destructive environmental forces such as earthquakes and winds. The new control algorithm, wavelet-hybrid feedback linear mean squared (LMS) algorithm, integrates a feedback control algorithm such as the LQR or LQG algorithm with the filtered-x LMS algorithm and utilizes a wavelet multi-resolution analysis for the low-pass filtering of external dynamic excitations. The goals are to achieve optimum control under external dynamic disturbances in real time and to overcome shortcomings of the existing feedback control algorithms and the filtered-x LMS algorithm.

The second objective of this book is to devise a new hybrid control system, hybrid damper-tuned liquid column damper (TLCD) system. The new hybrid control system, which combines passive and semi-active control systems, is intended to achieve increased reliability and maximum operability of the control system during power failure, and to eliminate the need for a large power requirement unlike other proposed hybrid control systems where active and passive systems are combined.

The great majority of papers published in the area of active structural vibration control deal with small or academic problems. Nearly a decade ago, the senior author wrote *Control, Optimization, and Smart Structures - High-Performance Bridges and Buildings of the Future* (Adeli and Saleh, 1999) where the control models were applied to large and realistic multistory building structures. The same philosophy is continued in the present book. The work presented in this book is based on some advanced mathematical concepts. The models are tested and their effectiveness is evaluated extensively on small problems for the sake of comparison with other methods and results reported in the literature. But they are also applied to realistic and large building and bridge structures. As such, the book demonstrates the applicability of the new smart technology developed in this work to large realistic civil structures.

1. 2. Overview of the book

In Chapter 2, major types of control systems — passive, active and semi-active control systems — are introduced. Among the passive control systems, supplementary damper, tuned mass damper (TMD), and tuned liquid column damper (TLCD) systems are described and their control performance is investigated using an 8-story shear building frame. Chapter 3 provides a primer on wavelets. Basic concepts are introduced and wavelet multiresolution analysis is described.

A method is presented for time-frequency signal analysis of earthquake records using Mexican hat wavelets in Chapter 4. The proposed signal processing methodology can be used to investigate the characteristics of accelerograms recorded on various types of sites and their effects on different types of structures.

Chapter 5 presents the classical feedback control algorithms. Their shortcomings are demonstrated using an active tuned mass damper system. The adaptive filtered-x Least Mean Square (LMS) control algorithm, based on the integration of the adaptive filter theory used for system identification in real time and the feedforward control approach, is described in Chapter 6. In Chapter 7, a hybrid feedback-LMS algorithm, which integrates a feedback control algorithm such as LQR and LQG algorithms and the filtered-x LMS algorithm, is presented. The hybrid feedback-LMS control algorithm is intended to achieve faster vibration suppression than the filtered-x LMS algorithm, and to be capable of suppressing vibrations over a range of input excitation frequencies unlike the classic feedback control algorithms whose control effectiveness decreases considerably when the frequency of the external disturbance differs from the fundamental frequency of the system.

In Chapter 8, the wavelet-hybrid feedback LMS algorithm is presented through judicious integration of the hybrid feedback-LMS algorithm and a wavelet low-pass filter. The wavelet low-pass filter is introduced for better stabilization of the FIR filter during adaptation when applying the algorithm to the control of civil structures against real environmental forces.

In Chapter 9, the hybrid damper-TLCD control model presented in Chapter 2 is used for control of responses of three dimensional (3D) irregular buildings under various seismic excitations. The equations of motion for the combined building and TLCD system are derived for multistory building structures with rigid floors and plan and elevation irregularities. Then, optimal control of 3D irregular buildings equipped with a hybrid damper-TLCD system is described. The wavelet-hybrid feedback LMS control algorithm, presented in Chapter 8, is applied to find the optimum control forces. Irregular buildings are particularly susceptible to strong ground motions. Two multistory moment-resisting building structures with vertical and plan irregularities are used to investigate the effectiveness of the new control system in controlling the seismic response of irregular buildings.

In Chapter 10, the effectiveness of both the semi-active TLCD system and the hybrid damper-TLCD control system, presented in Chapters 2 and 8, is further investigated for the control of wind-induced motion of highrise buildings. Simulation results are presented for a 76-story building benchmark control problem. The performances of semi-active TLCD and hybrid damper-TLCD control systems are compared with that of a sample ATMD system.

In Chapter 11, the wavelet-hybrid feedback LMS algorithm is used for vibration control of cable-stayed bridges under various seismic excitations. Its effectiveness is investigated through numerical simulation using a benchmark control problem. The performance of the new algorithm is compared with that of a sample LQG controller. Numerical simulations are performed to evaluate the sensitivity of the control model to modeling errors and verify its robustness. Finally, Chapter 11 provides some concluding remarks.

Chapter 2
Vibration Control of Structures

2. 1. Introduction

In this chapter, major types of control systems — passive, active and semi-active control systems — are introduced. Among the passive control systems, supplementary damper, tuned mass damper (TMD), and tuned liquid column damper (TLCD) systems are introduced and their control performance is investigated using an 8-story shear building.

2. 2. Passive control of structures

Passive control refers to systems that do not require an external power source. It includes base isolation, supplementary damper, and tuned mass damper (TMD) systems. A base isolation system [Figure 2.1(a)] attempts to reduce the response of structures subjected to seismic ground excitations by isolating the structure from the external seismic excitations. The seismic isolation system is usually applied to relatively massive buildings that are housing sensitive equipment such as computer centers, emergency operation centers, hospitals, and nuclear power plants, and to the rehabilitation of historic-landmark buildings such as the Los Angeles City Hall (Youssef et al., 2000) and the Ninth Circuit U.S. Court of Appeals building located in San Francisco (Mokha et al., 1996). The base isolation systems used in these applications are often large, heavy, and costly.

2.2.1. Supplementary damper devices

The supplementary damper system [Figure 2.1(b)] has been widely used for vibration suppression in general. In this system, mechanical devices increase the existing inherent damping of the structure and help dissipate the energy of the external excitation. The mechanical dampers in buildings are usually installed as part of their bracing system, such as diagonal or Chevron bracings [Figure 2.1(b)]. Examples include an 11-story steel building located in Sacramento, California (Miyamoto and Scholl, 1998) and the seismic upgrade of a 13-story concrete frame structure located in Los Angeles, California (Hanson and Soong, 2001). Supplementary dampers are sometimes used with other types of passive and/or active devices in order to maximize the suppression capacity (Youssef et al., 2000).

Supplementary dampers are grouped into two major categories; hysteretic devices and viscoelastic devices (Hanson and Soong, 2001). Hysteretic devices include metallic yielding and friction devices. They rely

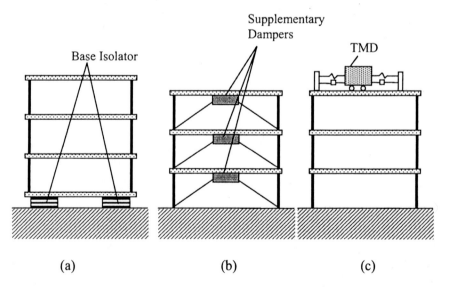

(a)   (b)   (c)

Figure 2.1 Three passive control devices: (a) base isolation system, (b) supplementary damper system, (c) TMD system

primarily on relative displacement for their energy dissipation. Energy dissipation of viscoelastic devices, in general, depends on their relative velocity as well as relative displacement.

2.2.2. Passive viscous fluid dampers

Among the viscoelastic devices, the viscous fluid device is one of the most widely used in practice recently. A typical viscous fluid damper is a cylindrical device containing incompressible silicon oil, where energy is dissipated as the oil passes through a small orifice. Unlike other viscoelastic devices, the viscous fluid damper does not introduce additional stiffness into the structure. Consequently, its energy dissipation depends only on the relative velocity. Because of this feature as well as simplicity of installation and small size, viscous fluid dampers have been applied to a number of real life structural applications (Soong and Constantinou, 1994; Miyamoto and Scholl, 1998). Additional detailed review and valuable information about supplementary damping devices can be found in a book by Hanson and Soong (2001).

Since the energy dissipation of a viscous fluid device depends only on its relative velocity, its output force can be expressed as a function of its relative velocity, $\dot{u}_v$, as follows:

$$f_v(t) = f_v[\dot{u}_v(t)] \tag{2.1}$$

For the orifice-controlled viscous fluid damper, its output force can be expressed as a power function of the relative velocity (Hanson and Soong, 2001) in the following form:

$$f_v(t) = c_v |\dot{u}_v(t)|^\beta \operatorname{sgn}(\dot{u}_v(t)) \tag{2.2}$$

where c_v is the generalized damping coefficient, sgn represents the sign function, and β is a coefficient in the range of 0.3 to 2.0. Values of β smaller than 1.0 are effective in reducing vibrations. For structures subjected to earthquake or wind loading, a value of one is often used. The value of $\beta = 1$ is used in this work. In that case, Eq. (2.2) is simplified as

$$f_v(t) = c_v \dot{u}_v \tag{2.3}$$

Then, the governing equation of motion for an m-DOF (degree of freedom) discrete structural system with multiple supplementary dampers subjected to external excitation is

$$M\ddot{u}(t) + C\dot{u}(t) + Ku(t) = B_v f_v(\dot{u}_v(t)) + E_e f_e(t) \tag{2.4}$$

where M, C, and K are m x m mass, damping, and stiffness matrices, respectively; $u(t) = m$ x 1 displacement vector; $f_v(t) = l$ x 1 supplementary damping force vector; $f_e(t) = r$ x 1 external dynamic force vector; B_v and E_e are m x l and m x r location matrices which define locations of the supplementary damping forces and the external excitations, respectively, $l =$ number of supplementary dampers, $r =$ dimension of external excitation, and t is the time.

The required damping capacity is determined based on the desired performance level, for example, desired lateral displacement of structures. Generally speaking, dampers with a larger damping coefficient and more dampers result in a more effective response reduction. However, depending on the flexibility/rigidity of a given structure and dynamic characteristics of external disturbance, acceleration and displacement may not always be decreased even when damping is increased. Feng and Shinozuka (1993) report that the increased damping applied to base-isolated bridges results in an increase in the absolute acceleration as well as relative displacements. A similar observation is reported by Sadek and Mohraz (1998) where the authors conclude that increasing damping in flexible structures (with fundamental period longer than 1.5 seconds) increases the acceleration response while decreasing the relative displacements. Therefore, the damping level of supplementary dampers needs to be chosen carefully considering the type of the structure.

In addition to the size of the supplementary damper defining the magnitude of the damping force, the locations and number of dampers need to be selected. In terms of the selection of number and locations, viscous fluid dampers provide great flexibility as one can choose from a range of a relatively large number of low-capacity dampers to a relatively small number of high-capacity dampers. For instance, the Taylor Damper Company provides a list of about 90 bridge and 3- to 67-story building structures, built or designed to be built, where orifice-controlled viscous fluid dampers with capacity ranging from 10 kN to 6700 kN have been used (http://www.taylordevices.com/3seismic.htm).

2.2.3. Tuned mass damper

A TMD system [Figure 2.1(c)] relies on the damping forces introduced through the inertia force of a secondary system attached to the main structure in order to reduce the response of the main structure. The secondary mass is designed to have dynamic characteristics that are closely related to those of the primary structure. The most important characteristics are the mass ratio of the secondary mass to the primary system, the frequency ratio of the two systems, and the damping ratio of the secondary system. By varying these three ratios, the frequency response function of the primary system can be modified so that the response of the primary system is reduced. Examples can be found in the John Hancock tower in Boston and the Citicorp Building in New York City (Housner et al., 1997).

In a TMD system, the secondary mass, which is usually made of concrete or steel, is attached to the main structure through a spring and a dashpot. The parameters of mass, spring, and dashpot are often tuned to the fundamental natural frequency of the structures so that the maximum response reduction occurs near that frequency. The drawback of this approach is that a TMD system provides protection against external dynamic disturbances with a frequency only in the vicinity of the natural frequency of the structure and not for a range of frequencies or bandwidth normally found in environmental forces. Moreover, to find the optimal values of parameters for a TMD system, the magnitude of the external excitation must be established *a priori*, which is not practical considering the variable nature of environmental forces. To overcome these shortcomings, active and semi-active TMD systems have been proposed where values of the parameters are changed based on the frequency and the amplitude of excitation in real time (Hrovat et al., 1983; Abe, 1996). Others have proposed the Multiple-TMD system where more than one TMD system is designed and distributed within the structure to cover a range of dominant frequencies (Kareem and Kline, 1995).

2.2.4. Tuned liquid column damper

More recently, the tuned liquid column damper (TLCD) has received the attention of researchers (Sakai et al., 1989; Kareem, 1994; Won et al., 1996; Yalla et al., 2001) as another type of secondary mass system (Figure 2.2). Similar to a TMD system, a TLCD system can reduce the response of the primary system by modifying its frequency response function. In a TLCD system, the secondary mass is liquid and damping forces are introduced through the motion of liquid in a U-shape tube container. When the same mass is used and other parameters are properly tuned, a TLCD system

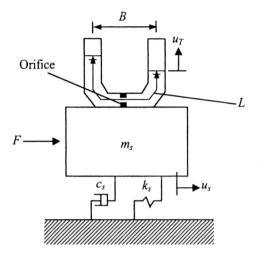

Figure 2.2 An SDOF system with a TLCD system

provides performance similar to a TMD system (Samali et al., 1998).

In addition to reducing building responses, a TLCD system provides several advantages over a TMD system as follows:

• The required level of damping can be readily achieved and controlled through the orifice/valve, making it suitable not only for passive control systems but also for semi-active control systems.

• When there are changes in the dynamic characteristics of the main structure after construction is completed or after the occurrence of an earthquake, the TLCD parameters (frequency and mass) can be easily tuned by adjusting the height of the liquid in the tube.

• The liquid in the system is easily mobilized at all levels of the structural motion, thereby eliminating the activation mechanism required in the conventional TMD system where a certain level of threshold excitation must be set.

• Water contained in the tube can be utilized as a secondary water source for an emergency such as fire.

- It provides configuration and space flexibilities as one can design one large tube or a group of smaller tubes.

Because of advantages, a growing number of bridge and building structures have been built with the TLCD system over the past decade or so. Examples include the Higash-Kobe cable-stayed bridge in Japan (Sakai et al., 1991), the 106.2-m high Hotel Cosima in Tokyo (Teramura and Yoshida, 1996), and the 194.4-m high Shin Yokohama Prince Hotel in Japan (Kareem, 1994). Recently, a TLCD system was used in the 48-story One Wall Centre, the tallest building in Vancouver, British Columbia. Two specially designed U-shaped tanks, each containing about 50,000 gallons (189 tons) of water, are installed in the tower's mechanical penthouse in order to lessen the lateral movement of the building against both earthquakes and strong winds. According to Fortner (2001), the TLCD system has saved at least 2 million dollars in construction costs compared to the conventional TMD system. This is because the TLCD system eliminates the installation of a pump station and a backup generator required for fire suppression. If the water is used for extinguishing a fire, the effectiveness of the TLCD system is reduced. When a fire occurs during an earthquake, the TLCD system will protect the structure during ground motions. The water in the TLCD system can then be used to extinguish any ensuing fire. In addition, the water tanks are used as heat sinks for the building's heat pump, and thick concrete water tank walls on the roof level act as outrigger walls.

Referring to Figure 2.2, a TLCD system is attached to an SDOF system; the equations of motion are (Sakai et al., 1989)

$$\begin{bmatrix} m_s + m_T & \alpha m_T \\ \alpha m_T & m_T \end{bmatrix} \begin{Bmatrix} \ddot{u}_s(t) \\ \ddot{u}_T(t) \end{Bmatrix} + \begin{bmatrix} c_s & 0 \\ 0 & \dfrac{\rho A \xi(t)|\dot{u}_T(t)|}{2} \end{bmatrix} \begin{Bmatrix} \dot{u}_s(t) \\ \dot{u}_T(t) \end{Bmatrix}$$

$$+ \begin{bmatrix} k_s & 0 \\ 0 & 2\rho A g \end{bmatrix} \begin{Bmatrix} u_s(t) \\ u_T(t) \end{Bmatrix} = \begin{Bmatrix} F(t) \\ 0 \end{Bmatrix} \tag{2.5}$$

where m_s, k_s, and c_s, are mass, stiffness, and damping coefficient of the SDOF primary system, respectively; $m_T = \rho A L$ is the mass of the liquid; $\alpha = B/L$ is the length ratio of the liquid tube; ρ, A, B, and L are the density, the cross-sectional area, the width and the length of the liquid tube, respectively; u_s is the horizontal displacement of the SDOF primary system; u_T is the vertical displacement of the liquid in the liquid column; $\xi(t)$ is the coefficient of head loss determined by the opening ratio (opening percentage) of the orifice at time t; and g is the gravitational acceleration. The second equation in Eq. (2.5) represents the nonlinear equation of the motion of the TLCD. The natural frequency of the TLCD can be obtained as

$$\omega_T = \sqrt{\frac{2\rho A g}{\rho A L}} = \sqrt{\frac{2g}{L}} \tag{2.6}$$

Equation (2.6) shows that the natural frequency of the TLCD system depends only on the length of the liquid tube. Analogous to TMD systems, tuning ratio, f, and the mass ratio, μ, of a TLCD system relative to the primary system are given by

$$f = \frac{\omega_T}{\omega_s} = \frac{\sqrt{2g/L}}{\omega_s} \tag{2.7}$$

$$\mu = \frac{m_T}{m_s} \tag{2.8}$$

where ω_s is the natural frequency of the primary system. From Eq. (2.5), the equivalent damping ratio $\zeta(t)$ of the TLCD system at time t can be expressed as

$$\zeta(t) = \frac{\xi(t)}{4\sqrt{2gL}}|\dot{u}_T(t)| \tag{2.9}$$

When the primary system is controlled passively, the head loss coefficient $\xi(t)$ has a constant value. Yet, damping ratio of the TLCD system is also dependent on the velocity of liquid as noted in Eq. (2.9). The relationship between the value of the head loss coefficient and the orifice opening ratio is estimated experimentally (Balendra et al., 1995) and tabulated in the literature (Blevins, 1984).

Equation (2.5) is now expanded for an MDOF (multi-degree of freedom) system attached to a TLCD subjected to an earthquake ground acceleration $\ddot{x}_g$ as follows:

$$\begin{bmatrix} M+M' & M_{ST} \\ M_{TS} & m_T \end{bmatrix} \begin{Bmatrix} \ddot{u}(t) \\ \ddot{u}_T(t) \end{Bmatrix} + \begin{bmatrix} C & [0]_{mx1} \\ [0]_{1xm} & c(t) \end{bmatrix} \begin{Bmatrix} \dot{u}(t) \\ \dot{u}_T(t) \end{Bmatrix} \\ + \begin{bmatrix} K & [0]_{mx1} \\ [0]_{1xm} & 2\rho Ag \end{bmatrix} \begin{Bmatrix} u(t) \\ u_T(t) \end{Bmatrix} = -\begin{Bmatrix} MJ \\ m_T \end{Bmatrix} \ddot{x}_g(t) \tag{2.10}$$

where mass coupling matrices M_{ST} and M_{TS} and the mass contribution of TLCD to the primary system mass matrix represented by M' are

$$M_{ST} = \begin{bmatrix} [0]_{(m-1)x1} \\ \alpha m_T \end{bmatrix} \tag{2.11}$$

$$M_{TS} = M_{ST}^T \tag{2.12}$$

$$M' = \begin{bmatrix} [\boldsymbol{0}]_{(m-1)\times(m-1)} & [\boldsymbol{0}]_{(m-1)\times 1} \\ [\boldsymbol{0}]_{1\times(m-1)} & m_T \end{bmatrix} \tag{2.13}$$

and

$$c(t) = \frac{\rho A \xi(t)}{2} |\dot{u}_T(t)| \tag{2.14}$$

in which $J = m \times 1$ is a column vector with all elements equal to one.

In order to compare the effectiveness of a TLCD to that of a TMD, an 8-story shear building frame presented in Yang (1982) is examined here (Figure 2.3). This particular structure is chosen because the same example has been used as a test example by a number of other researchers (Yang et al., 1987; Soong, 1990; Spencer et al., 1994). The structural properties are: floor mass = 345.6 tons, elastic stiffness of each story = 3.404 x 105 KN/m, and internal damping coefficient of each story = 2,937 kN-sec/m. The damping coefficient corresponds to a 2 percent damping for the fundamental vibration mode of the entire structure.

Three simulated earthquake ground accelerations used in Yang et al. (1987) and Spencer et al. (1994) are employed (denoted by EQ-I, EQ-II, and EQ-III). They are stochastic signals with a Kanai-Tajimi spectral density defined by

$$S(\omega) = S_0 \left[\frac{4\zeta_g^2 \omega_g^2 \omega^2 + \omega_g^4}{(\omega^2 - \omega_g^2)^2 + 4\zeta_g^2 \omega_g^2 \omega^2} \right] \tag{2.15}$$

where parameters ζ_g, ω_g, and S_0 represent the soil damping property, the dominant frequency of the ground motion, and amplitude intensity of the motion, respectively. The values of these parameters depend on the characteristics and intensity of the ground acceleration in a particular geological location. The values of parameters ζ_g and ω_g for three simulated

18

earthquake ground accelerations are presented in Table 2.1. The value of S_0 is set to be 4.5 x 10^{-4} m^2/sec^3.

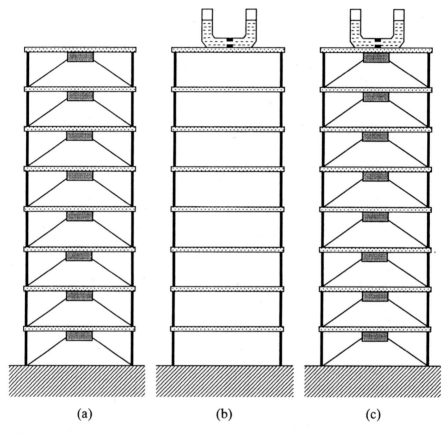

(a)   (b)   (c)

Figure 2.3 Eight-story shear building frame: (a) with passive/semi-active supplementary damper system, (b) with passive/semi-active TLCD system, (c) with hybrid system

Table 2.1 Parameters of simulated earthquake ground acceleration (Spencer et al. 1994)

Parameters	EQ-I	EQ-II	EQ-III
ζ_g	0.65	0.064	0.317
ω_g (rad/sec.)	18.85	31.12	10.516

For EQ-I, the following time envelope function, $\tau(t)$ is used to specify the shape and duration of the earthquake ground acceleration

$$\tau(t) = \begin{cases} (t/t_1)^2 & 0 \le t < t_1 \\ 1 & \text{for } t_1 \le t < t_2 \\ \exp[-c(t-t_2)] & t > t_2 \end{cases} \qquad (2.16)$$

where $t_1 = 3$ sec, $t_2 = 13$ sec, and $c = 0.26$ sec-1 are used, following Yang et al. (1987). This particular earthquake ground acceleration, shown in Figure 2.4, is used in this chapter to present the response time histories and the maximum responses of the example structure with various control systems. EQ-II and EQ-III ground accelerations simulate approximately the 1955 San Jose N59E and 1952 Kern County N90E earthquakes, respectively. These earthquake ground accelerations as well as EQ-I are used to measure the root mean square (RMS) responses of the structure.

Figure 2.5 shows time histories of the top floor displacement of the 8-story frame of Figure 2.3 for the uncontrolled, and passively controlled

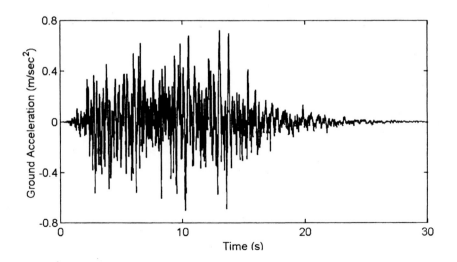

Figure 2.4 Simulated earthquake ground acceleration, EQ-I

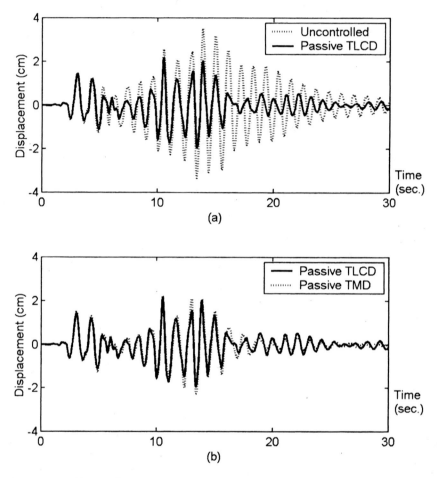

Figure 2.5 Time history of top floor displacement: (a) uncontrolled and passive TLCD controlled responses, (b) passive TMD and passive TLCD controlled responses

TLCD and TMD systems subjected to EQ-I. In order to compare the effectiveness of the TLCD system with that of the TMD system, the same tuning and mass ratios of $f = 0.98$ and $\mu = 0.02$ provided by Yang (1982) for a TMD system are used. The mass ratio, μ, is the ratio of the mass of the TLCD to the generalized mass associated with the first mode of the primary MDOF system. Optimum head loss coefficient of 1.78 is used following

Yalla and Kareem (2000). Figure 2.5(b) shows the responses of TLCD and TMD systems in terms of reducing the response are not significantly different when similar design parameters are optimized and used for a given earthquake ground acceleration. Therefore, the TLCD system is preferred over the conventional TMD system because of its practical advantages noted earlier and similar effectiveness.

As in TMD systems, however, the effectiveness of a TLCD system depends on proper tuning of design parameters. Most important design parameters include mass ratio μ, tuning ratio f, and head loss coefficient ξ. These parameters are usually obtained such that the TLCD system minimizes the response in a root mean square sense for a given external excitation. Tuning ratios near but less than one and larger mass ratios generally result in a more effective control of structures. Sadek et al. (1998) and Won et al. (1996) conclude that the optimal head loss coefficient ξ increases as the amplitude of excitation decreases and the mass ratio increases. Like TMD systems, however, optimum values of these parameters are obtained only for any given external excitations with fixed frequency bandwidth and amplitude. In other words, these values are optimal only for the design excitation and not any other external excitation. This shortcoming can be overcome by utilizing semi-active or active control strategies.

2. 3. Active control of structures

In order to improve the performance of passive control systems, active control systems have been proposed where sensors measure the motions of the structure and actuators and a feedback control strategy exert counteracting forces to compensate for the effect of external excitations (Saleh and Adeli, 1994; Adeli and Saleh, 1997, 1998 and 1999; Saleh and Adeli, 1998; Jiang and Adeli, 2008a and b; Christenson et al., 2003). A shortcoming of active control of structures is its dependency on a large power requirement for the control system. An active control system will not operate when a strong earthquake causes the failure of the electric power system unless there is a large properly operating backup battery system.

2. 4. Semi-active control of structures

Semi-active control strategies have been proposed by researchers to increase the overall reliability as well as the efficacy of the control system (Housner et al., 1997). Semi-active control systems are physically similar to passive control systems but computationally similar to active control systems. Developed from passive control devices, semi-active control devices are designed to operate with a very small power (e.g., a battery) thus eliminating the need for a large external electric power source. They control the response of the structure by actively changing the properties of controllers when power is supplied, but behave like passive control systems when the power source is cut off or when there is a computer system failure. As such, semi-active control systems provide a more reliable and stable way of controlling structures compared with active control systems.

There is another strategy to overcome the vulnerability of active control systems, called hybrid control, where two distinct systems are employed together. Traditionally, an active control system is used in conjunction with a passive control system (Soong and Reinhorn, 1993; Lee-Glauser et al., 1997). When there is power (normally electric power) the two systems work simultaneously. When the external power fails the passive control system still works, thus reducing the response of the structure at least to some extent even after the active control system stops functioning. The shortcoming of this approach is that in the event of power failure during a catastrophic or maximum probable earthquake only one half of the earthquake resistant system is available and safety of the structural system is not guaranteed.

2.4.1. Semi-active viscous fluid damper

Orifice-controlled viscous fluid dampers can be relatively easily modified into semi-active control devices requiring a small power only. This is achieved simply by modulating the size of the opening in the orifice. Developed originally in military, aerospace, and automotive industries, the application of semi-active dampers in civil structures has received the attention of researchers during the past decade (Symans and Constantinou, 1999). Recently, an actual building equipped with semi-active variable dampers was built in Japan (Kurata et al., 1999).

In a semi-active control system, the value of the damping coefficient cannot be negative. It is practically bounded in the range of a minimum value, c_{vmin}, and a maximum value, c_{vmax}. Also, the control forces are constrained to be in the opposite directions of the velocities of the corresponding dampers in order to improve their efficacy. Consequently, the value of the damping coefficient for damper i at time t is regulated in accordance with the following constraint:

$$c_v^i(t) = \begin{cases} c_{v\min} & c_v^{i*}(t) < c_{v\min} \text{ or } f_v(t)\dot{u}_v^i(t) \geq 0 \\ c_v^{i*}(t) & \\ c_{v\max} & c_v^{i*}(t) > c_{v\max} \end{cases} \tag{2.17}$$

where $c_v^{i*}(t)$ is the optimal damping coefficient for damper i at time t obtained from the control algorithm adopted.

As in passive damper systems, the effectiveness of the semi-active damper also depends on the flexibility of the structure. Symans and Constantinou (1997) tested a variable semi-active fluid damper for a three-story frame with fundamental frequency of 1.8 Hz analytically and experimentally. They report the same effectiveness for variable semi-active dampers as passive dampers in reducing the structural response. Research on effectiveness of semi-active dampers was also carried by Sadek and Mohraz (1998) for SDOF systems having fundamental period in the range of 0.2 to 3.0 second. They conclude that efficiency of variable semi-active dampers is questionable for rigid structures (with fundamental period less than 1.5 seconds) compared with passive dampers. Even for flexible structures such as base-isolated structures, they report that compared with passive damper systems, semi-active systems improve acceleration response suppression to some extent without any additional suppression of displacement response. In some cases, the displacement is even slightly increased. A similar observation is made by Singh and Matheu (1997) who concluded that semi-active damper systems yield no significant benefit over the passive damper system.

In order to compare the effectiveness of a semi-active viscous fluid damper to that of a passive damper, the same 8-story shear-building frame presented in previous section is examined (Figure 2.3). The same damper is used in every story. The damping coefficient for each supplementary damper (c_v in Eq. 2.9) is chosen such that roughly the same level of reduction of top floor displacement is obtained as the TMD system provided in Yang (1982), resulting in a value of 3,500 kN-sec/m (20 kip-sec/in.), which is well within the practical range of commercially available viscous fluid dampers. This addition of dampers increases the damping in the fundamental mode of the controlled structure to about 5.5 percent.

Figure 2.6 shows time histories of the top floor displacement of the 8-story frame of Figure 2.3(b) for three cases: uncontrolled structure, and passively and semi-actively controlled structure with supplementary dampers subjected to EQ-I. The corresponding maximum accelerations, shear forces, and displacements at different stories are shown in Figure 2.7. In order to find the optimal damping coefficient in Eq. (2.17), the LQR-based semi-active control algorithm provided by Sadek and Mohraz (1998) is used. The value of c_{vmax} used for the semi-active control system is 3,500 kN-sec/m (20 kips-sec/in.), the same value used for the damping coefficient of the passive

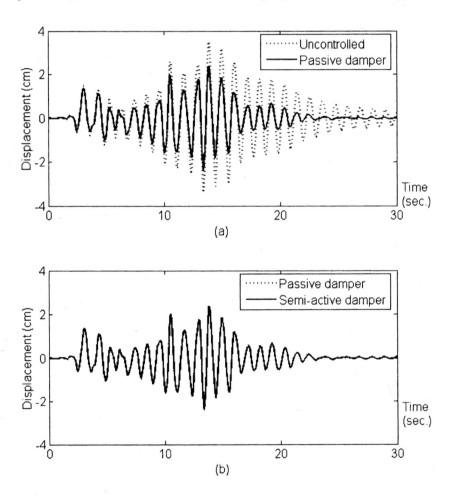

Figure 2.6 Time histories of the top floor displacement of the 8-story frame of Figure 2.3(b) for three cases: uncontrolled structure, and passively and semi-actively controlled structure with supplementary dampers subjected to EQ-I

26

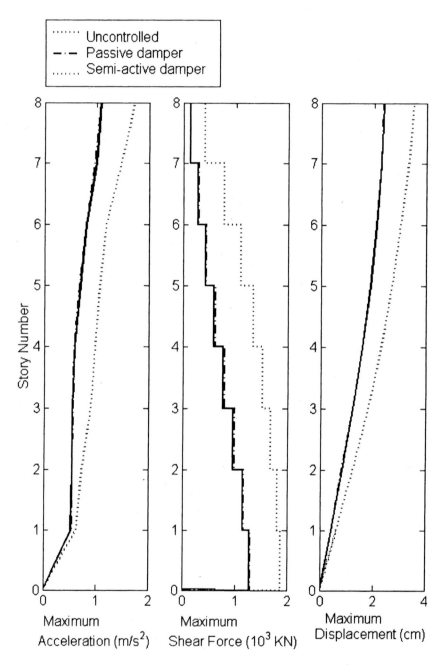

Figure 2.7 Maximum accelerations, shear forces, and displacements for uncontrolled, passively controlled, and semi-actively controlled structure subjected to EQ-I

system. The value of c_{vmin} for the semi-active control system is set to zero. Results of this investigation and Figures 2.6 and 2.7 indicate that the semi-active viscous fluid damper system provides no noticeable improvement in reducing the displacement and shear force response over the less complicated and less costly passive system while increasing the acceleration responses slightly. Similar observations are found in an experimental study by Symans and Constantinou (1997) using ER (electrorheological) dampers, and analytical studies by Singh and Matheu (1997) and Sadek and Mohraz (1998) using variable dampers.

2.4.2. Semi-active TLCD system

If the head loss coefficient $\xi(t)$ in Eq. (2.9) can be changed by a controllable orifice, then the passive damping force is transformed into an active force which controls the response of the structure. Equation (2.10) can be re-written as

$$\begin{bmatrix} M + M' & M_{ST} \\ M_{TS} & m_T \end{bmatrix}\begin{Bmatrix} \ddot{u}(t) \\ \ddot{u}_T(t) \end{Bmatrix} + \begin{bmatrix} C & [0]_{m \times 1} \\ [0]_{1 \times m} & 0 \end{bmatrix}\begin{Bmatrix} \dot{u}(t) \\ \dot{u}_T(t) \end{Bmatrix} + \begin{bmatrix} K & [0]_{m \times 1} \\ [0]_{1 \times m} & 2\rho A g \end{bmatrix}\begin{Bmatrix} u(t) \\ u_T(t) \end{Bmatrix}$$

$$= -\begin{Bmatrix} MJ \\ m_T \end{Bmatrix}\ddot{x}_g(t) + \begin{Bmatrix} 0 \\ 1 \end{Bmatrix}f_c(t)$$

(2.18)

where

$$f_c(t) = -c(t)\dot{u}_T(t) = -\frac{\rho A \xi(t)|\dot{u}_T(t)|}{2}\dot{u}_T(t)$$

(2.19)

Similar to the semi-active fluid damper system, the value of head loss coefficient is regulated in accordance with the semi-active control law expressed as

$$\xi(t) = \begin{cases} \xi_{\min} & \xi^*(t) < \xi_{\min} \text{ or } f_c(t)\ddot{u}_T(t) \geq 0 \\ \xi^*(t) & \\ \xi_{\max} & \xi^*(t) > \xi_{\max} \end{cases} \qquad (2.20)$$

where $\xi_{\max}$ and $\xi_{\min}$ are, respectively, the maximum and minimum limits of head loss coefficient and $\xi^*(t)$ is the optimal head loss coefficient at time t obtained from the control algorithm adopted. In practice, the value of $\xi_{\max}$ for the semi-active TLCD system is set to be greater than the optimal value of ξ obtained for the passive TLCD system in order to cover a range of amplitude and frequency of excitations. This is because the optimal value of ξ for the passive TLCD is determined in the root mean square sense and design earthquake ground excitation cannot be known *a priori*.

Figure 2.8(a) presents the time histories of top floor displacements for uncontrolled and semi-active TLCD systems shown in Figure 2.3(a) subjected to EQ-I. Figure 2.8(b) presents the time histories of top floor displacements for passive and semi-active TLCD systems subjected to EQ-I. The corresponding maximum accelerations, shear forces, and displacements per story are shown in Figure 2.9. In these numerical simulations of passive and semi-active TLCD systems, a value of 15 is used for $\xi_{\max}$. The value of $\xi_{\min}$ is set to be zero because the head loss coefficient cannot have a negative value practically. As observed in Figures 2.8 and 2.9, the semi-active TLCD system can yield significant improvement for response reduction over the passive TLCD system unlike semi-active dampers (Figures 2.6 and 2.7).

2. 5. Hybrid control of structures

In the previous section, the effectiveness of the semi-active TLCD system over the passive TLCD system was demonstrated. By optimally adjusting the head loss coefficient, the semi-active TLCD system can achieve a significant improvement over the passive TLCD system. However, the performance of either semi-active or passive TLCD system is bounded by mass and tuning ratios of a liquid tube. Even though a TLCD system with a larger mass ratio may yield more effective response reductions, the larger mass ratio may increase the stiffness requirement of the primary structure in order to support the larger mass at the top. This may result in an uneconomical design. Also, values of the mass and tuning ratios are limited by the space and length

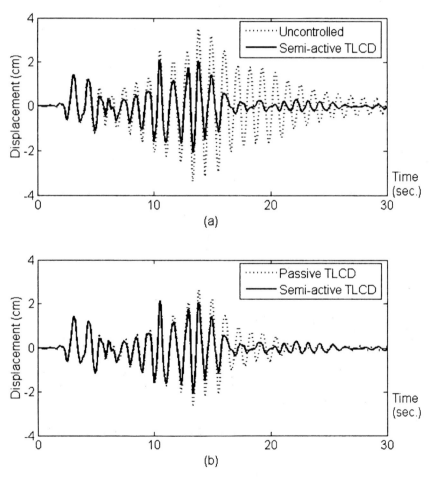

Figure 2.8 Time history of top floor displacements: (a) uncontrolled and semi-active TLCD controlled responses, (b) passive and semi-active TLCD controlled responses

30

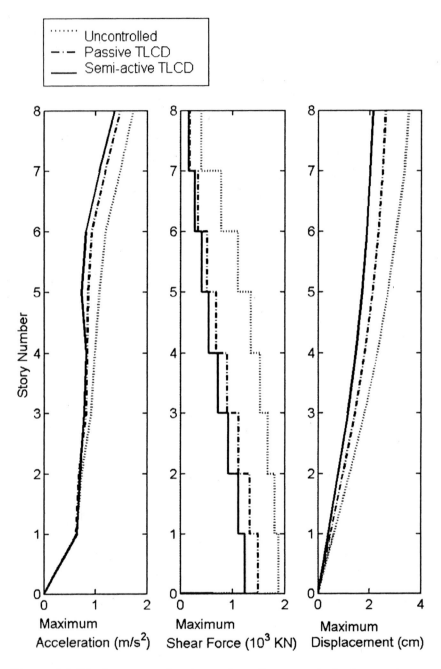

Figure 2.9 Maximum accelerations, shear forces, and displacements per story

available for the TLCD system.

In this section, the semi-active TLCD system is integrated with passive viscous fluid passive damper devices in order to overcome the shortcomings of the semi-active TLCD system and enhance its reliability and vibration reduction capability. Viscous fluid dampers are used because they do not introduce any additional stiffness and can provide any desired damping force. Moreover, a passive damper system is inherently reliable because it does not depend on an external electric power source. The entire hybrid damper-TLCD control system can operate on very small power, e.g., a battery, without having to rely on a large external electric power. This elimination of the need for a large power requirement makes the proposed hybrid control system more reliable than other hybrid control systems where active and passive systems are combined.

2.5.1. Steps involved in the design and implementation of the hybrid damper-TLCD system

The main steps involved in the design and implementation of the proposed hybrid damper-TLCD system are summarized in this section.

1. Determine the design parameters of TLCD: mass ratio μ and tuning ratio f as discussed earlier. These parameters should also be determined based on the trade-off between the desired performance level and practicality. Larger mass ratios may produce more effective response reduction, but the cost, space, and weight of mass may prevent the use of large mass ratios. The tuning ratio depends only on the length of the liquid tube (Eq. 2.7), and the tube may have an irregular shape depending on the required tube length and available space.

2. Select between the continuous and on-off type orifice/valve controller based on cost and practical implementation considerations. In the former, which is more effective, the opening ratio of the orifice can be changed continuously. In the latter, which is usually less expensive, the opening ratio of the orifice can have just two values, a minimum and a maximum value.

3. Determine the maximum value of the head loss coefficient ξ_{max}. In order to cover a range of excitation amplitude and frequency bandwidth, as a rule of thumb, the value of ξ_{max} for the semi-active system used in the proposed hybrid system should be greater than the value of constant ξ for the passive TLCD system, which is based on the statistical RMS value computed for a given external excitation. However, very large values of ξ_{max} may not be practical. In the case of a power and/or computer system failure, the opening ratio of the orifice cannot be changed and is generally set equal to its minimum or maximum value. This also limits the upper value for ξ_{max}. The power and/or computer system failures most probably are encountered during a strong earthquake when a large value of ξ_{max} may have an adverse effect because larger magnitudes of excitation require smaller values of head loss coefficient when the semi-active TLCD system acts like a passive TLCD system. If the required performance level is not achieved, add passive dampers as described in the next step.

4. Determine the required damping ratios and configuration of supplementary dampers based on performance level requirements. The selection of the damping ratio and damper configuration should be based on the trade-off between the desired response reduction and

other factors such as cost, available damper capacity, and architectural considerations.

2.5.2. Evaluation of the effectiveness of the hybrid damper-TLCD system

In order to evaluate the effectiveness of the proposed hybrid damper-TLCD system under various seismic excitations, numerical simulations are performed for the 8-story frame shown in Figure 2.3(c) using the three simulated earthquake ground accelerations discussed earlier in this chapter (Eq. 2.15 and Table 2.1). For the supplementary passive dampers in the hybrid damper-TLCD system, the same uniform damper configuration and damping coefficient are used as those used for the simulation of the passive viscous fluid dampers. Also, the same design parameters used for the semi-active TLCD system are used here in order to compare the effectiveness of the proposed system.

Time histories of top floor displacements for the 8-story frame of Figure 2.3(c) subjected to EQ-I are presented in Figure 2.10. Figure 2.10(a) presents time histories of top floor displacement for the passive damper and hybrid damper-TLCD systems. Figure 2.10(b) presents time histories of top floor displacement for the semi-active TLCD and hybrid damper-TLCD systems. The corresponding maximum accelerations, shear forces, and displacements per story are presented in Figure 2.11.

The maximum responses of the top floor and maximum base shear forces of the structure subjected to EQ-I are summarized in Table 2.2. Maximum top story displacement of the proposed hybrid damper-TLCD system is 25% and 17% less than the corresponding values for the passive damper and semi-active TLCD systems, respectively. Maximum top story acceleration of the hybrid damper-TLCD system is 12% and 30% less than the corresponding value for the passive damper and semi-active TLCD

34

systems, respectively. Maximum shear force (base shear) of the hybrid damper-TLCD system is 22% and 14% less than the corresponding value for the passive damper and semi-active TLCD systems, respectively.

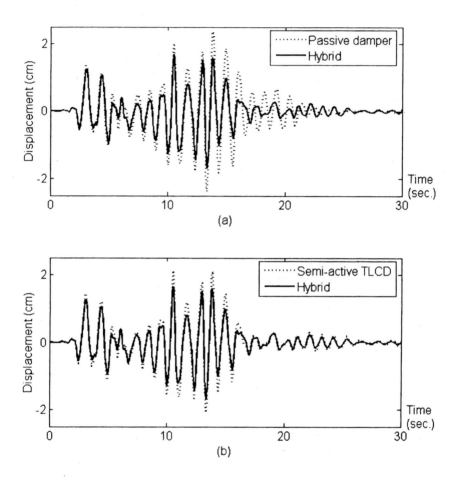

Figure 2.10 Time histories of top floor displacement for the 8-story frame of Figure 2.3(c): (a) passive damper and hybrid system, (b) semi-active TLCD and hybrid system

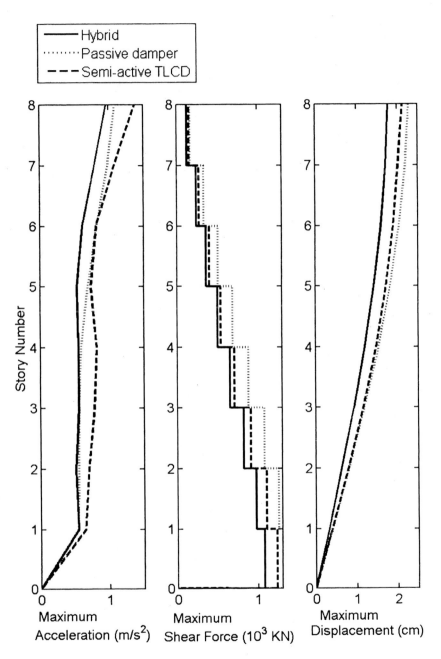

Figure 2.11 Maximum accelerations, shear forces, and displacements per story

Table 2.2 Maximum responses of top floor and maximum base shear forces of the structure subjected to EQ-I

Response	Uncontrolled	Passive damper	Semi-active TLCD	Hybrid damper-TLCD
Displacements max$|u_8|$ (cm)	3.52	2.39	2.13	1.77
Accelerations max$|\ddot{u}_8|$ (m/sec^2)	1.74	1.09	1.37	0.96
Maximum base shear force (x 10^3 KN)	1.88	1.37	1.24	1.07

Note: u_8= displacement of the 8th (top) floor; $\ddot{u}_8$ = acceleration of the 8th (top) floor

Table 2.3 RMS responses of top floor of the structure subjected to simulated earthquake ground accelerations, EQ-I, EQ-II, and EQ-III

Response	Uncontrolled	Passive damper	Semi-active TLCD	Hybrid damper-TLCD
EQ-I				
RMS (u_8) (cm)	1.12	0.67	0.59	0.49
RMS ($\ddot{u}_8$) (m/sec^2)	0.40	0.24	0.25	0.19
EQ-II				
RMS (u_8) (cm)	1.30	0.88	0.86	0.74
RMS ($\ddot{u}_8$) (m/sec^2)	0.58	0.34	0.47	0.30
EQ-III				
RMS (u_8) (cm)	1.97	1.45	1.42	1.28
RMS ($\ddot{u}_8$) (m/sec^2)	0.76	0.59	0.63	0.57

Table 2.3 presents RMS acceleration and displacement responses of the top floor subjected to all three simulated earthquake ground accelerations, EQ-I, EQ-II, and EQ-III. As for the maximum responses summarized in

Table 2.3, RMS responses of the hybrid damper-TLCD system are consistently lower than the corresponding responses of both passive damper and semi-active TLCD systems. From Tables 2.2 and 2.3, it is concluded that the hybrid damper-TLCD system can effectively reduce the responses of structures subjected to different earthquake ground accelerations.

Figure 2.12 shows the time history of the top floor displacements for

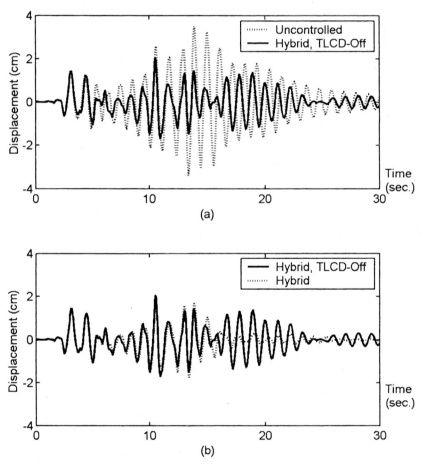

Figure 2.12 Time histories of top floor displacements: (a) uncontrolled system and hybrid system when the semi-active TLCD controller is not functioning fully (TLCD-Off), (b) hybrid controlled system and hybrid system when the semi-active TLCD controller is not functioning fully (TLCD-Off)

the hybrid damper-TLCD system subjected to EQ-I when the semi-active TLCD controller is functioning as a passive system only due to power or computer failure (denoted as TLCD-Off in the figure) along with the response when the hybrid system is functioning fully. For the TLCD-Off case, the value of the head loss coefficient is fixed at ξ_{min}, assuming a passive rather than a semi-active TLCD system. Even though the performance of the TLCD-Off case does not match that of the whole hybrid system, especially in the second half of the simulation [Figure 2.12(b)], significant response reduction is still achieved compared with the uncontrolled response [Figure 2.12(a)]. Similar results are obtained for the TLCD-Off case with the value of the head loss coefficient fixed at ξ_{max} but are not presented here for the sake of brevity. The results demonstrate that the proposed hybrid damper-TLCD system is stable and robust in terms of power or computer failure.

2. 6. Concluding Remarks

For both supplementary damper and TLCD systems, damping is achieved and damping forces are controlled through an orifice/valve, making them suitable not only for passive control systems but also for semi-active control systems. However, it is shown that the performance improvement of semi-active viscous fluid damper systems over the less complicated and less costly passive damper systems is not always guaranteed depending on the flexibility of the structure. On the other hand, a semi-active TLCD system can reduce the response significantly compared with a passive TLCD system. This can be explained by the fact that the head loss coefficients are modified continuously on-line based on the frequency and magnitude of external excitations.

A new hybrid control model was presented by combining supplementary passive damper and semi-active TLCD systems. It is found that the new model is effective in significantly reducing the response of an MDOF system under various seismic excitations. Also, it is shown that the hybrid control system provides increased reliability and maximum operability during normal operations as well as a power or computer failure. The proposed system eliminates the need for a large power requirement, unlike other proposed hybrid control systems where active and passive systems are combined.

Chapter 3
Wavelets

3. 1. What is a wavelet?

The wavelet transform is a relatively recent mathematical transformation method (Daubechies, 1992; Adeli and Samant, 2000; Adeli and Karim, 2000; Samant and Adeli, 2000 and 2001; Karim and Adeli, 2002a and b; Wu and Adeli, 2001; Karim and Adeli, 2003; Adeli et al., 2003; Ghosh-Dastidar and Adeli, 2003; Zhou and Adeli, 2003a and b; Jiang and Adeli, 2003; Adeli and Ghosh-Dastidar, 2004; Adeli and Kim, 2004; Jiang and Adeli, 2004; Sirca and Adeli, 2004; Adeli and Karim, 2005; Jiang and Adeli, 2005a and b; Adeli and Jiang, 2006; Ghosh-Dastidar and Adeli, 2006; Adeli et al., 2007; Jiang and Adeli, 2007; Ghosh-Dastidar et al., 2007; Jiang and Adeli, 2008a and b). The original signal is transformed into a different domain where a more comprehensive analysis and processing becomes possible. Similar to conventional transform methods such as the Fourier transform, the wavelet transform represents the original signal as a linear combination of basis functions. But, instead of breaking down a signal into a series of basis functions over an infinite range, the original signal is broken down into a series of basis functions that are localized in both time and frequency. Due to locality in both time and frequency domains of its basis function, the wavelet transform provides an effective way of processing signals characterized by time-varying nonstationary frequency contents (Newland, 1993).

3. 2. Types of wavelets

If $\psi(t)$ is the basis wavelet function, called a mother wavelet, the members of family are defined as (Rao and Bopardikar, 1998)

$$\psi_{a,b}(t) = \frac{1}{a}\psi\left(\frac{t-b}{a}\right) \tag{3.1}$$

where a and b are real numbers and indicate the scaling and translation of the mother wavelet, respectively. The scaling parameter, a, represents the frequency content of the wavelet. The translation parameter, b, represents the location of wavelet in time. Thus, in contrast to the Fourier transform, the basis function of the wavelet transform retains the time locality as well as frequency locality.

In general, the family of wavelets defined by Eq. (3.1) need not be orthogonal. But, orthogonal wavelets require a substantially fewer number of operations compared with non-orthogonal wavelets, and therefore orthogonal wavelets are used in this work, with the exception of Chaper 4. The wavelet set $\{\psi_{a,b}\}$ forms an orthogonal system if (Daubechies, 1992; Meyer, 1993)

$$\left\langle \psi_{a,b}, \psi_{a,c} \right\rangle = \int \psi_{a,b}(t)\psi_{a,c}(t)dt = 0 , \; b \neq c \tag{3.2}$$

where $\left\langle \cdot,\cdot \right\rangle$ represents the inner product. Denoting the number of data to be transformed by N, the wavelet transform using orthogonal wavelets requires only $O(N)$ operations in contrast to $O(N\log N)$ operations needed for the fast Fourier transform resulting in much faster transformation (Newland, 1993).

The basis wavelet functions satisfy the following prescribed conditions:

- Continuity;
- A zero mean amplitude (the integral of the function is equal to zero).

This implies that a wavelet function has at least some oscillations;

• The wavelet function takes either null values outside a given real domain of R (it has a finite duration or energy, and consequently known as compactly supported) or the function approaches zero quickly as the independent variable approaches infinity. This property prevents the propagation of any local transient signal features through time indefinitely; and

• Relatively less energy for lower frequencies compared with that for higher frequencies

There are other properties that classify wavelet functions into different categories. For example,

• Orthogonality. Orthogonal wavelets are one type of compactly supported wavelet functions where the same basis function is used for both decomposition and reconstruction.

• Biorthogonality. Biorthogonal wavelets are another type of finite duration wavelet functions where there is a pair of basis wavelet functions that are dual to each other, that is, if $\psi_{a,b}(t)$ is the decomposition wavelet basis, then its dual reconstruction or synthesis basis is $\widetilde{\psi}_{j,k}(t)$ where

$$< \psi_{a,b}(t), \widetilde{\psi}_{j,k}(t) >= \delta_{a,b}\delta_{j,k} \qquad a, b, j, k \in Z \qquad (3.3)$$

in which Z is the domain of integers, and $\delta_{a,b}$ and $\delta_{j,k}$ are Kronecker delta symbols ($\delta_{a,b} = 1$ if $a=b$ and $\delta_{a,b} = 0$ if $a \neq b$);

• Admissibility. It ensures the perfect reconstruction of the original signal from the transformed wavelet coefficients;

- Symmetry or antisymmetry. It refers to the shape of the wavelet;
- Number of vanishing moments of the basis wavelet function. This property is useful for data compression applications; and
- Regularity. It refers to the degree of differentiability of the wavelet function, which is essential when smooth representation of signals is needed.

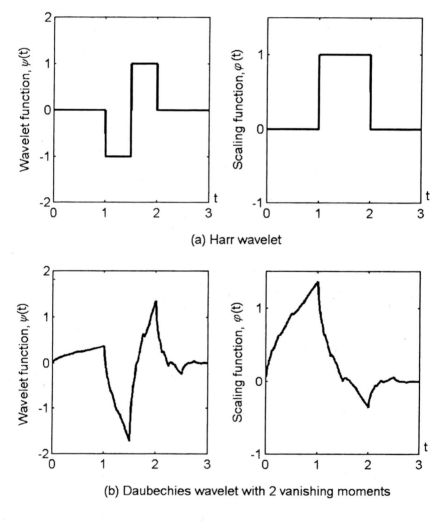

(a) Harr wavelet

(b) Daubechies wavelet with 2 vanishing moments

Figure 3.1 Wavelet and scaling functions for Harr wavelet and Daubechies wavelet with 2 vanishing moments

Figure 3.1 shows two examples of orthogonal wavelets: (a) Harr wavelet and (b) Daubechies wavelet with 2 vanishing moments. A wavelet with i-vanishing moments means that "there exists a certain function θ such

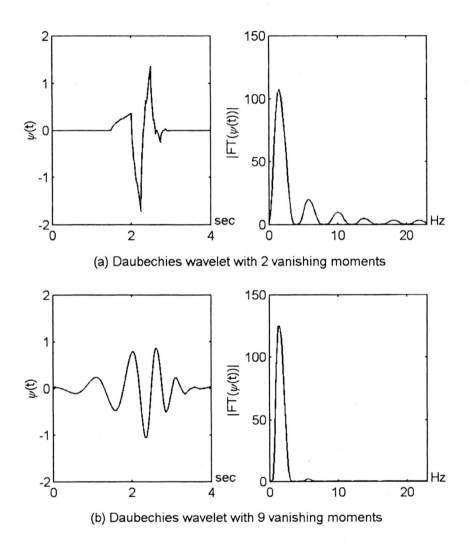

(a) Daubechies wavelet with 2 vanishing moments

(b) Daubechies wavelet with 9 vanishing moments

Figure 3.2 Daubechies wavelet functions and their Fourier transforms (denoted by FT): (a) with 2 vanishing moments, (b) with 9 vanishing moments

that the wavelet can be written as the ith order derivative of θ' (Mallat, 1989). In this sense, the Harr wavelet can be regarded as a Daubechies wavelet with 1 vanishing moment.

Figure 3.2 shows Daubechies wavelet functions with 2 and 9 vanishing moments along with their Fourier transforms (FT) which show their frequency contents. The larger the vanishing moments, the more computation is required since there are more coefficients involved. Figure 3.2 demonstrates clearly that a Daubechies wavelet function with larger vanishing moments provides better frequency locality, whereas a Daubechies wavelet function with smaller vanishing moments shows better time locality. Also, comparing Figures 3.1 and 3.2, it is seen that a wavelet function with larger vanishing moments is smoother than another one with smaller vanishing moments.

3. 3. Multiresolution analysis

If the input is defined in a discrete domain and the dyadic dilation is applied, Eq. (3.1) can be expressed as

$$\psi_{j,k}(n) = 2^{-j/2}\psi\left(2^{-j}t - n\right) \tag{3.4}$$

where $j, k \in Z$. In addition to the wavelet function, the family members of the basic scaling function are defined by scaling and translation as

$$\varphi_{j,k}(n) = 2^{-j/2}\varphi\left(2^{-j}n - k\right) \tag{3.5}$$

The relationship between the wavelet function $\psi(n)$ and the scaling functions $\varphi(n)$ are defined such that the set of functions $\psi_{j,k}(n)$ span the difference W_j (wavelet function space) between the scaling function spaces, V_j, while the scaling function spaces are spanned by the various scales of the scaling function as follows:

$$V_j = \underset{k}{\text{Span}}\{\varphi_{j,k}(n)\} \qquad (3.6)$$

$$W_j = V_{j+1} \ominus V_j \qquad (3.7)$$

where $\ominus$ represents a direct subtraction.

The wavelet and scaling functions, $\psi(n)$ and $\varphi(n)$, constitute the key elements of the multiresolution analysis (MRA) (Mallat, 1989) which can be employed for filtering purposes. The MRA is formulated with a nesting of the spanned spaces as

$$\cdots \subset V_{-2} \subset V_{-1} \subset V_0 \subset V_1 \subset V_2 \subset \cdots \subset L^2(R) \qquad (3.8)$$

where $V_{-\infty} = \{0\}$ (null space) and $V_\infty = L^2(R)$ is the space of all square integrable functions. W_j's, $j = -\infty,\ldots, \infty$, are orthogonal to each other because of their definition (Eq. 3.7) and relationships among V_j's, $j = -\infty,\ldots, \infty$ (Eq. 3.8). Based on the definition of V_j, we can write the following natural scaling condition for any function $f(n)$:

$$f(n) \in V_j \quad \leftrightarrow \quad f(2n) \in V_{j+1} \qquad (3.9)$$

If Eqs. (3.6)-(3.9) hold, then there exists a set of functions $\psi_{j,k}$ such that $\psi_{j,k}$ $(k \in Z)$ spans W_j which is the orthogonal complement of the spaces V_j and V_{j+1}. More specifically, if $\{\varphi_{0,k}\}$ spans V_0 then $\{\psi_{0,k}\}$ spans W_0 such that

$$V_1 = V_0 \oplus W_0 \qquad (3.10)$$

and

$$L^2(R) = \cdots \oplus W_{-2} \oplus W_{-1} \oplus W_0 \oplus W_1 \oplus W_2 \oplus \cdots \qquad (3.11)$$

where $\oplus$ represents a direct sum. This means that by starting with a representation of a function belonging to a coarse subspace, higher detail or resolution can be obtained by adding spaces spanned by $\psi_{j,k}$ at a higher resolution (i.e., given by the next higher value of j).

A discrete input signal, $x(n)$, can be represented as a combination of wavelet and scaling functions as follows:

$$x(n) = \sum_k c_{j_0,k} \varphi_{j_0,k}(n) + \sum_k \sum_{j=j_0} d_{j,k} \psi_{j,k}(n) \qquad (3.12)$$

where the first term is a coarse resolution at scale j_0 and the second term adds details of increasing resolutions. Equation (3.12) can also be viewed as the time-frequency decomposition of $x(n)$ where the second term provides the frequency and time breakdowns of the signal.

From the nesting of the spaces spanned by scaling functions represented by Eq. (3.8) and the relationship between the spaces spanned by wavelet functions and those spanned by scaling functions expressed by Eqs. (3.7) and (3.10), we can write (Mallat, 1989)

$$\varphi(t) = \sum_k h_0[k] \varphi(2t - k) \qquad k \in Z \qquad (3.13)$$

$$\psi(t) = \sum_k h_1[k] \varphi(2t - k) \qquad k \in Z \qquad (3.14)$$

where h_0 and h_1 are filter coefficients. The filter coefficients are obtained by solving Eqs. (3.13) and (3.14). Unique and exact solutions for Eqs. (3.13) and (3.14) exist only when k is equal to 2, 4, and 6, corresponding to the Daubechies wavelet with 1 vanishing moment (or Haar wavelet), 2 vanishing moments [Figure 3.1(b)], and 3 vanishing moments, respectively. For example, the Daubechies wavelet with 2 vanishing moments has the following h_0 coefficients: $\dfrac{(1+\sqrt{3})}{4}, \dfrac{(3+\sqrt{3})}{4}, \dfrac{(3-\sqrt{3})}{4}, \dfrac{(1-\sqrt{3})}{4}$. The h_1 coefficients can be found by the following equation (Burrus et al., 1998):

$$h_1[n] = (-1)^n h_0[3-n] \tag{3.15}$$

When k is greater than 6, no unique solution exists, and thereby the filter coefficients are obtained numerically by adjusting the coefficients iteratively until the resulting wavelet has desirable decomposition and resolution properties. Illustrative examples on the iterative numerical solution for Eqs. (3.13) and (3.14) can be found in Newland (1993).

For a given set of h_0 and h_1 coefficients, the wavelet decomposition can be performed by a two-band filter bank using the time-reversed filters $h_0[-n]$ (low-pass filter) and $h_1[-n]$ (high-pass filter) followed by down-sampling by a factor of 2. The down-sampling by a factor of 2 takes a signal $x(n)$ as input and produces output of $x(2n)$. In practice, this down-sampling is achieved by taking every other term of an input signal. A two-band multi-level filter bank (or filter tree) is shown in Figure 3.3.

Figures 3.5 and 3.6 show an example of multiresolution analysis for an example block signal generated using Matlab (2000) and presented in Figure

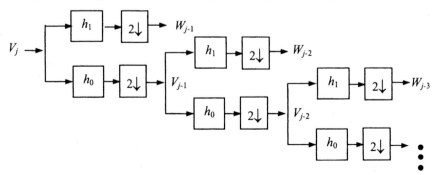

Figure 3.3 A two-band multi-level filter tree

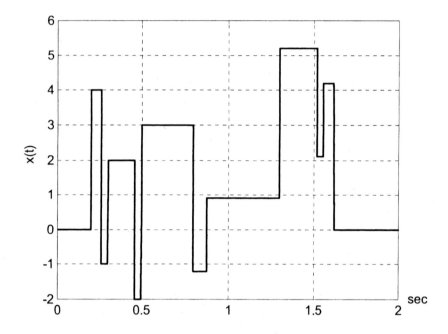

Figure 3.4 An example block signal generated using Matlab

3.4. Figure 3.5 shows the approximations of the block signal in various scaling function spaces V_j using a Daubechies scaling function with 2 vanishing moments. This figure illustrates how the approximations progress: higher and more accurate resolutions are achieved at spaces with higher scaling functions. The projection of the original block signal onto the highest scaling function space, V_{10}, yields the original signal itself exactly. Figure 3.6 illustrates the individual wavelet decomposition by showing the components of the signal that exist in the wavelet function spaces W_j at different scales j. The relationship between scaling function spaces, V_j, and wavelet function spaces, W_j, represented by Eq. (3.9) can be verified in Figures 3.5 and 3.6. For example, $V_2 = V_1 \oplus W_1$, which in single dimension means the simple addition of projections of the original signal onto spaces V_1 and W_1 yields the projection of the signal onto space V_2. The projection of the original block

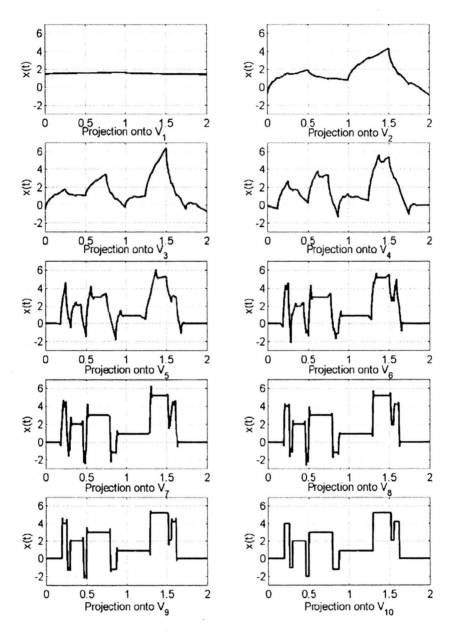

Figure 3.5 Projection of the block signal shown in Figure 3.4 onto *V* spaces using a Daubechies scaling function with 2 vanishing moments

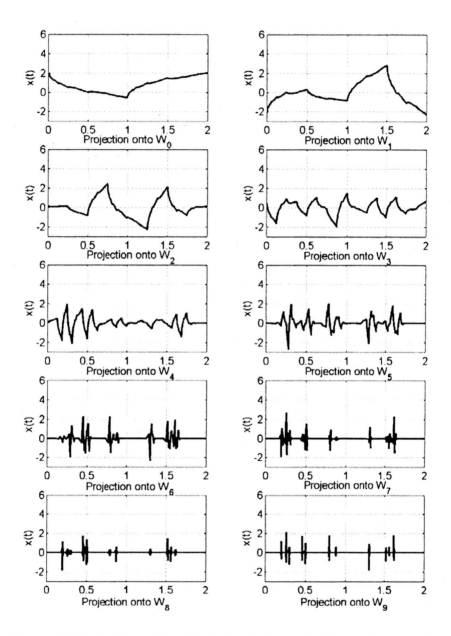

Figure 3.6 Projection of the block signal shown in Figure 3.4 onto W spaces using a Daubechies wavelet function with 2 vanishing moments

signal onto the highest wavelet function space W9 shows the locations of the edges of the original signal quite accurately. In other words, high-frequency content of the signal over the time axis can be represented accurately with the highest wavelet function space. The example presented in Figures 3.4 to 3.6 shows that the wavelet transform provides an effective way of processing signals in time and frequency domains simultaneously.

Chapter 4

Time-Frequency Signal Analysis

of Earthquake Records[1]

4. 1. Introduction

Wavelet analysis originally was developed by mathematicians and seismologists working on seismic signal analysis at about the same time (Goupillaud, et. al. 1984; Grossmann and Morlet, 1984; Daubechies, 1988). Seismologists' interest in and contribution to wavelets stem from the fact that earthquake signals (seismograms or accelerograms) are non-stationary transient time sequences. Recently, a number of articles have been published on earthquake signal processing using wavelet transforms, including detection of the arrival time of P or primary wave and S or secondary wave (Oonincx, 1999), prediction of future earthquakes (Alperovich and Zheludev, 1998; Lyubushin, 1999), and strong ground motion synthesis (Iyama and Kuwamura, 1999).

However, little research has been reported in the literature on the use of wavelet transform to analyze seismic data from a structural engineering point of view and to study the dynamic behavior of structures under seismic loading. Basu and Gupta (1997) present wavelet-based stochastic analysis of a linear multi-degree of freedom system under earthquake loading. A closed

---

[1] This chapter is based on the article: Zhou, Z. and Adeli, H. (2003), "Time-Frequency Signal Analysis of Earthquake Records Using Mexican Hat Wavelets", *Computer-Aided Civil and Infrastructure Engineering*, Vol. 18, No. 5, pp. 379-389, and is reproduced by permission of the publisher, Wiley-Blackwell.

form solution is obtained for the instantaneous power spectral density function using time and frequency localization of the wavelet transform. The wavelet used in the analysis is a slightly modified version of the Littlewood-Paley (L-P) wavelet (Newland, 1993). They conclude that the formulation can estimate different ordered peak responses of a seven-story shear building frame under ground motion reasonably well. Iyama and Kuwamura (1999) apply wavelet transform to ground motions from the viewpoint of energy input to structures. They discuss simulation of ground accelerations by wavelet inverse transform where accelerations are simulated based on a given history of instantaneous energy input specified for each frequency. Through simulation of a single-degree-of-freedom (SDOF) system subjected to the 1995 Hyogoken-Nanbu earthquake ground motions (with magnitude M=6.9) recorded at different stations, they conclude that the wavelet coefficients represent energy responses of structures.

Recently, Sirca and Adeli (2004) developed a new method of generating artificial earthquake accelerograms through integration of artificial neural networks (Adeli and Hung, 1995) and wavelets. A counterpropagation (CPN) neural network model (Adeli and Park, 1995a, 1998) is developed for generating artificial accelerograms from any given design spectrum such as the International Building Code (IBC) design spectrum (IBC, 2000). In order to improve the efficiency of the model, the CPN network is modified with the addition of the wavelet transform as a data compression tool to create a new CPN-wavelet network. Wavelets have also been used recently for signal processing of traffic data (Adeli and Samant, 2000; Adeli and Karim, 2000; Samant and Adeli, 2000; Karim and Adeli, 2002a and b).

In this chapter, a method is presented for time-frequency signal analysis of earthquake records using Mexican hat wavelets. Ground motions in earthquakes are postulated as a sequence of simple penny-shaped ruptures at different locations along a fault line and occurring at different times. The single point source displacement of ground motion is idealized by a Guassian function. For the purpose of signal analysis of accelerograms, the ground motion record generated by a simple penny-shaped rupture is used to form the basis wavelet function. The result of the signal processing of an accelerogram is presented in the form of a scalogram using the coefficients of the continuous Mexican hat wavelet transform to describe the signal energy in the time-scale domain. The presented signal processing methodology can be used to investigate the characteristics of accelerograms recorded on various types of sites and their effects on different types of structures.

4. 2. Continuous wavelet transform (CWT)

Let $f(t)$ be a square integrable function of time, t. The continuous wavelet transform of $f(t)$ is defined as (Chui, 1992)

$$W_{a,b} = \int_{-\infty}^{\infty} f(t) \frac{1}{\sqrt{|a|}} \psi * (\frac{t-b}{a}) dt \qquad (4.1)$$

where $a, b \in R$, $a \neq 0$, the star symbol "*" denotes the complex conjugation, and the wavelet function is defined as

$$\psi_{a,b}(t) = \frac{1}{\sqrt{|a|}} \psi(\frac{t-b}{a}) \qquad (4.2)$$

Equation (4.1) can be expressed as

$$W_{a,b} = \int_{-\infty}^{\infty} f(t) \psi_{a,b}^*(t) dt \qquad (4.3)$$

The factor $\dfrac{1}{\sqrt{|a|}}$ is used to normalize the energy so that it stays at the same level for different values of a and b; that is

$$\int_{-\infty}^{\infty} \left| \psi_{a,b}(t) \right|^2 dt = \int_{-\infty}^{\infty} \left| \psi(t) \right|^2 dt \tag{4.4}$$

The wavelet function $\psi_{a,b}(t)$ is expanded in time (or space) when a is increased, and displaced in time (or space) when b is varied. Therefore, a is called the scaling parameter which captures the local frequency content and b is called the translation parameter which localizes the wavelet basis function at time $t=b$ and its vicinity.

To implement CWT, the signal $f(t)$ is first sampled at discrete points on the time axis and then the set of scaling parameters a is chosen to achieve an appropriate range of frequency resolution. The set of translation parameters b is usually taken at the same points where the original signal $f(t)$ is sampled. After the parameters a and b are chosen, the basic wavelet, also called the mother wavelet, is dilated or compressed by the scaling factor a to produce a family of wavelets $\psi_{a,b}(t)$. The wavelets $\psi_{a,b}(t)$ are multiplied by $f(t)$ at different scales a and different translations b. The CWT coefficients $W_{a,b}$ are then obtained by summing the products, which indicate the correlation between the signal and the wavelet functions $\psi_{a,b}(t)$. As a result, at high frequencies, a good time resolution is achieved whereas at low frequencies, a good frequency resolution is obtained.

The original time domain signal can be reconstructed through the inverse wavelet transform (Daubechies, 1992)

$$f(t) = \frac{1}{2\pi C_\psi} \int_{-\infty}^{\infty} \int_{0}^{\infty} \frac{W_{a,b}}{a^2} \psi_{a,b}(t)\, da\, db \tag{4.5}$$

where

$$C_\psi = \int_0^\infty \frac{|\hat{\psi}(\omega)|^2}{|\omega|} d\omega < \infty \tag{4.6}$$

In Eq. (4.6), the hat sign "^" indicates $\hat{\psi}(\omega)$, a function of frequency (ω), is the Fourier transform of the basic wavelet function $\psi(t)$ and is given by

$$\hat{\psi}(\omega) = \frac{1}{\sqrt{2\pi}} \int_{-\infty}^\infty \psi(t) e^{-i\omega t} dt \tag{4.7}$$

The frequency content or spectrum information of the original signal $f(t)$ can also be obtained by computing its Fourier transform:

$$\hat{f}(\omega) = \frac{1}{\sqrt{2\pi}} \int_{-\infty}^\infty f(t) e^{-i\omega t} dt \tag{4.8}$$

In the Fourier transform of the original signal (Eq. 4.8), complex exponential or infinite sinusoid functions are used as basis functions. These infinite basis functions are suitable for extracting frequency information from periodic, non-transient signals. The Fourier transform, and in particular, the fast Fourier transform (FFT), has gained widespread acceptance during the past century in signal processing. However, the frequency spectrum of a signal as a result of the Fourier transform is not localized in time because of the infinite sinusoid basis functions. This implies that the Fourier coefficients of a signal are determined by the entire signal support. Consequently, if additional data are added over time, the Fourier transform coefficients will change. Any local behavior of a signal cannot be easily traced from its Fourier transformation.

In contrast, wavelet transform (Eq. 4.1) is a more suitable and powerful tool for analyzing transient signals, since both frequency (scales) and time information can be obtained. Long time intervals (corresponding to larger values of a) are used for more precise low-frequency information and shorter regions (corresponding to smaller values of a) for the time locality of high-frequency information. Furthermore, if the basis wavelet function in Eq. (4.2) is compactly supported, then the frequency information obtained from the wavelet transform is localized in time. Therefore, for transient signals the wavelet analysis is superior to Fourier transform due to its multi-resolution features.

The short time or windowed Fourier transform (SFT) (also known as Gabor transform, Gabor, 1946) is another time-frequency analysis method based on Fourier transform. In SFT, time and frequency information is localized by a uniform time window for all frequencies. In contrast, the wavelet transform adapts the window size according to the frequency. At high frequencies, fine resolution is obtained and at low frequencies, long windows are used to encompass those frequency contents. Therefore, the wavelet transform is also superior to SFT.

4. 3. Ground motions as a sequence of penny-shaped ruptures at different locations along the fault line

Most strong earthquakes of interest to structural engineers are caused by a sudden rupture or slip of a geological fault. An earthquake generated from a single point rupture can be described ideally as a penny-shaped crack located on the hypocenter (Housner, 1970). When the stress in the area inside the crack zone exceeds the rupture point, strain is released with the rupture and a displacement occurs between the two sides of the rupture as idealized in Figure 4.1.

A displacement wave is created by this rupture and propagates radially from the source. Vibrations are produced when this displacement wave reaches the earth's surface at the site of a structure. The general shape of such a simple single-source displacement wave is shown in Figure 4.2. In this work, we idealize the single-source displacement of ground motion by a Guassian function in the following form:

$$d(t) = e^{-\frac{1}{2}t^2} \tag{4.9}$$

Such a simple displacement wave has in fact been recorded at seismographic stations. There have been a number of small earthquakes with primarily one displacement wave similar to that shown in Figure 4.2. The first such simple-source ground motion was observed in the Port Hueneme earthquake on March 18, 1957 (with magnitude M=4.7), as shown in Figure 4.3.

Most earthquakes of engineering significance, however, are generated by a more complicated source mechanism and their ground motion appears more complex as noted, for example, in the time histories of the El

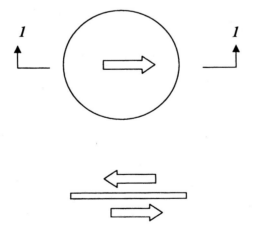

Section 1-1

Figure 4.1 Penny-shaped rupture to model single source earthquakes

Centro earthquake of 1940 (M=6.9) shown in Figure 4.4. It is postulated that ground motions in such earthquakes are produced by a sequence of the simple penny-shaped ruptures at different locations along a fault line. These simple ruptures occur at different locations and time, resulting in a series of simple records combined to create a complicated earthquake record such as that of the El Centro earthquake. The existence of a single-displacement earthquake wave supports the hypothesis of the decomposition of earthquakes into many single source ground motions of the kind shown in Figure 4.2.

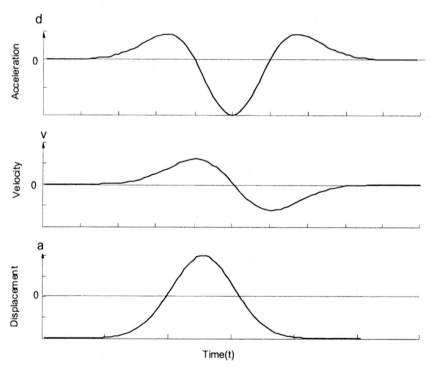

Figure 4.2 Idealized ground motion wave generated by a single source earthquake

4. 4. Selection of the basis wavelet function

For the purpose of signal analysis of accelerograms, the ground motion record generated by a simple penny-shaped rupture is used as the basis to form the mother wavelet. In this research, the selection of the most appropriate wavelet basis function is based on the following considerations:

- Since the mother wavelet should characterize the acceleration wave generated from a single rupture source, its initial and final values should approach zero. Consequently, the wavelet basis function has to be compactly supported with a finite duration, or nearly compactly supported.

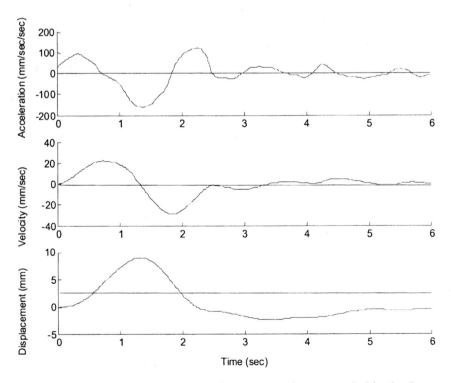

Figure 4.3 East-west component of ground motions recorded in the Port Hueneme earthquake (March 18, 1957)

- The goal of the wavelet decomposition of the earthquake accelerogram is to extract time-frequency features from the signal. As such, no reconstruction of the original accelerogram from the transformed wavelet coefficients is required. Therefore, the orthogonality or biorthogonality properties are not required.

- The acceleration generated by an ideal single displacement wave has a symmetric shape (Figure 4.2). Orthogonal wavelets are asymmetric in general (Daubechies, 1992) and therefore not suitable for our application. A nearly compactly supported Mexican hat wavelet is one wavelet with a symmetric shape as shown in Figure 4.5.

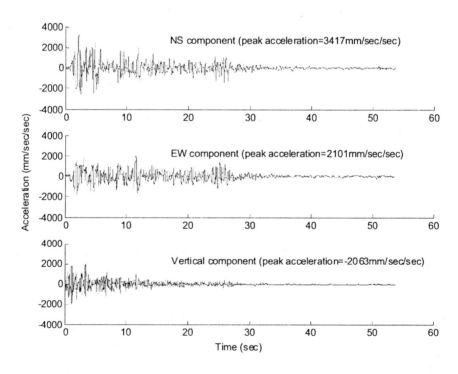

Figure 4.4 El Centro earthquake accelerograms (May 18, 1940)

- Since in response spectrum analysis a second order structural vibrations differential equation has to be solved, a wavelet with an analytical expression is preferred because it is more amenable to mathematical manipulation.

On the basis of the aforementioned considerations as well as a study of the shape of the acceleration record for a single point rupture (Figure 4.2), the Mexican hat wavelet (Figure 4.5) (Daubechies, 1992) is found in this research to be the most appropriate mother wavelet in the proposed method for time-frequency signal analysis of accelerograms. The Mexican hat wavelet can be described analytically by taking the second derivative of the Gaussian function, defined by Eq. (4.9),

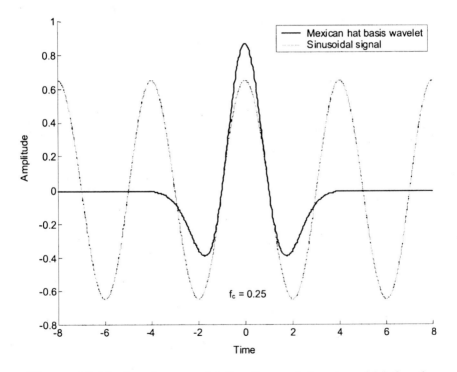

Figure 4.5 Mexican hat wavelet function and the sinusoidal function with a frequency equal to the center frequency of the wavelet

$$\psi(t) = (1 - t^2)e^{-t^2/2} \tag{4.10}$$

If Eq. (4.10) is normalized so that its norm becomes one (Daubechies, 1992), we obtain the following equation for the mother wavelet to be used in the proposed earthquake signal analysis:

$$\psi(t) = \frac{2}{\sqrt{3}}\pi^{-1/4}(1 - t^2)e^{-t^2/2} \tag{4.11}$$

4. 5. Representing earthquake acceleration signals by wavelet scalograms

In this section, we show how an earthquake accelerogram can be represented in the form of a scalogram using the coefficients of the continuous Mexican hat wavelet transform to describe the signal energy in the time-scale domain. The results of wavelet transform of a time series data (in this case, an accelerogram record), $f(t)$, are presented on a two-dimensional time-scale scalogram as a function of two variables, time and frequency. The horizontal axis represents the time or the translation parameter b. The vertical axis represents the frequency or the scaling parameter a.

The dominant frequency of a wavelet is called the center frequency of the wavelet, f_c (in Hz). It is one of the characteristics of any given wavelet function. The center frequency for the Mexican hat wavelet (Figure 4.5) is 0.25. In Figure 4.5, a sine function of the same frequency as f_c is plotted along with the Mexican hat wavelet for the sake of comparison.

A pseudo frequency is defined corresponding to scale a, f_a, (in Hz) as a function of the center frequency of wavelet, f_c, in the following way (Abry, 1997):

$$f_a = \frac{f_s f_c}{a} \tag{4.12}$$

where f_s is the sampling frequency of the original signal. The inverse of the pseudo frequency is the pseudo period for any given scale a. There is a linear relationship between the pseudo-period and the scale a (representing the frequency of the signal), as noted in Eq. (4.12) and shown in Figure 4.6. Most building structures have a natural period of less than 10 seconds with the exception of very tall (super highrise building) and slender structures. Therefore, as Figure 4.6 indicates, it is sufficient to use scales 1 to 128, corresponding to the pseudo periods of 0.08 seconds to 10 seconds, respectively.

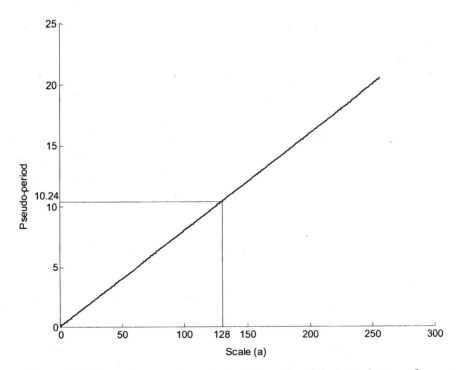

Figure 4.6 The scale-pseudo-period relationship of the wavelet transform using the Mexican hat wavelet at a sampling period of 0.02 sec.

68

The scalogram for the north-south component of the 1940 El Centro earthquake (Figure 4.4) using the Mexican hat wavelet decomposition is shown in Figure 4.7. Figure 4.8 shows the same results in a three-dimensional space. The scalogram for the east-west component of the same earthquake using the Mexican hat wavelet decomposition is shown in Figure 4.9. Figure 4.10 shows the same results in a three-dimensional space. Figure 4.11 shows the scalogram for the vertical component of the 1940 El Centro earthquake using the Mexican hat wavelet decomposition. Figure 4.12 shows the same results in a three-dimensional space. There is an inverse relationship between the wavelet scaling parameter a and the frequency of the signal. The

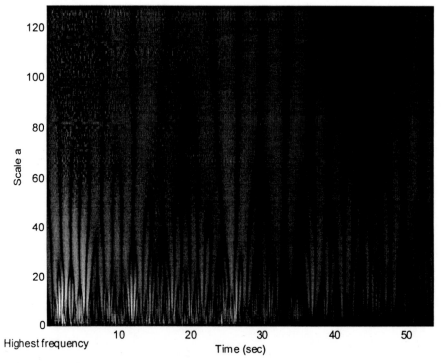

Figure 4.7 Absolute values of the wavelet coefficients for the 1940 El Centro earthquake (north-south component, scales = 1 to 128)

brighter portions in Figures 4.7, 4.9, and 4.11 indicate higher absolute values of the wavelet coefficients. These figures as well as Figures 4.8, 4.10, and 4.12 show which frequencies have the largest magnitude at any given time. Figures 4.11 and 4.12 show that the wavelet coefficients of the vertical component are more regularly distributed in both time and frequency, i.e., there is no systematic change in time across the frequencies and no change in frequencies across time. However, the magnitude of the vertical component is often much smaller and therefore attracts less attention in structural dynamic analysis.

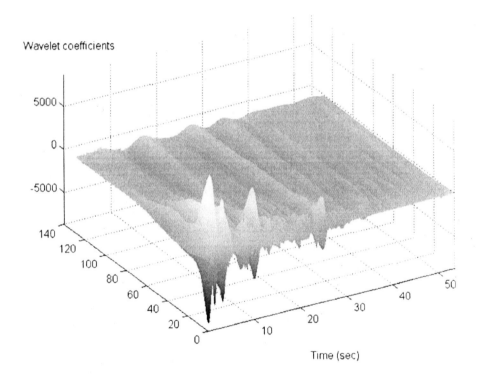

Figure 4.8 Three-dimensional surface plots of the wavelet coefficients for the 1940 El Centro earthquake (north-south component, scales = 1 to 128)

Figure 4.13 shows the north-south and east-west components of the ground accelerations due to the 1994 Northridge earthquake (M = 6.7). Figures 4.14 and 4.15 show the scalograms for these records. Scalorogramsfor the two horizontal components of the acceleration in the El Centro and Northridge earthquakes have similar time-frequency characteristics and evolution features. Similar observations were made for several other strong ground motion records.

Figure 4.16 shows one horizontal component of the ground acceleration of the 1971 San Fernando earthquake (M = 6.6). Figure 4.17 shows the scalogram for this record.

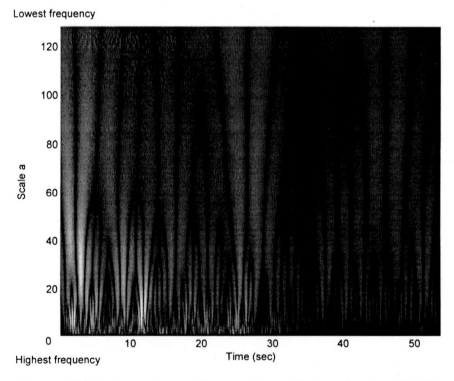

Figure 4.9 Absolute values of the wavelet coefficients for the 1940 El Centro earthquake (east-west component, scales = 1 to 128)

4. 6. Concluding remarks

In this chapter, a method was presented for time-frequency signal analysis of earthquake records using Mexican hat wavelets assuming that ground motions in earthquakes are produced by a sequence of simple penny-shaped ruptures at different locations along a fault line and occurring at different times. The result of the signal processing of an accelerogram is presented in the form of a scalogram. Scalograms were presented for several earthquake records.

Wavelet-based scalograms are an efficient way of obtaining time-frequency insight not readily obtained in other signal processing approaches. In a sense, they provide a microscopic time-frequency image of earthquake

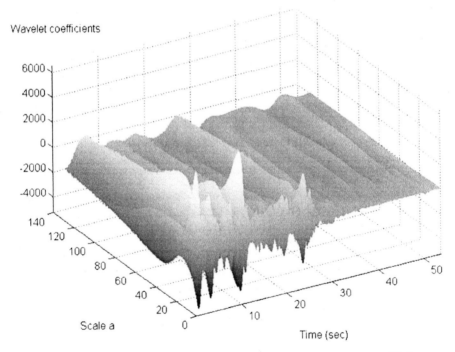

Figure 4.10 Three-dimensional surface plots of the wavelet coefficients for the 1940 El Centro earthquake (east-west component, scales = 1 to 128)

signals. For example, the scalograms of the horizontal ground accelerations presented in this chapter (Figures 4.7, 4.9, 4.14, 4.15, and 4.17) show that high-frequency contents are dominant mostly in the early stage of the motion. As time goes on lower frequency contents become more dominant. In Figures 4.14 and 4.17, the high absolute values of the wavelet coefficients move from the lower left in the early stage of the motion to the upper values in the mid-stage of the earthquake. This happens to be true for most accelerograms recorded on alluvium soils. This type of ground motion is harmful to most structures because the fundamental period of the structure usually increases during an earthquake due to the occurrence of cracks, damage to nonstructural elements, and loosening of the structural connections.

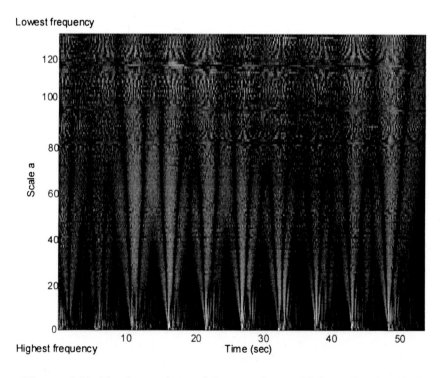

Figure 4.11 Absolute values of the wavelet coefficients for the 1940 El Centro earthquake (vertical component)

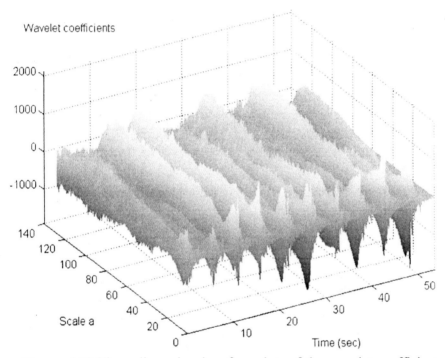

Figure 4.12 Three-dimensional surface plots of the wavelet coefficients for the 1940 El Centro earthquake (vertical component)

74

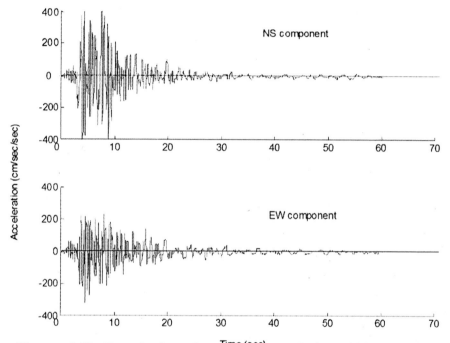

Figure 4.13 Two horizontal components of the 1994 Northridge earthquake

Lowest frequency

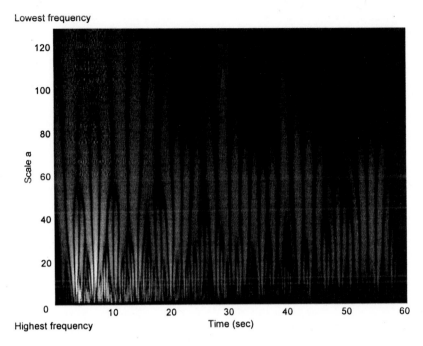

Highest frequency

Figure 4.14 Absolute values of the wavelet coefficients for the 1994 Northridge earthquake (north-south component, scales = 1 to

Lowest frequency

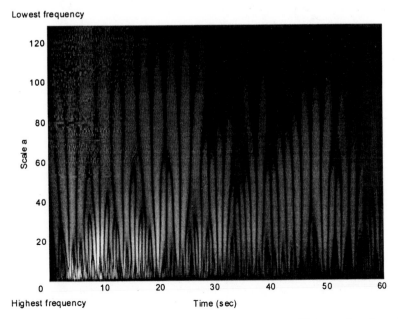

Highest frequency

Figure 4.15 Absolute values of the wavelet coefficients for the 1994 Northridge earthquake (east-west component, scales = 1

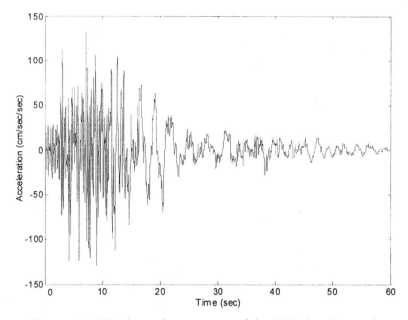

Figure 4.16 Horizontal component of the 1971 San Fernando

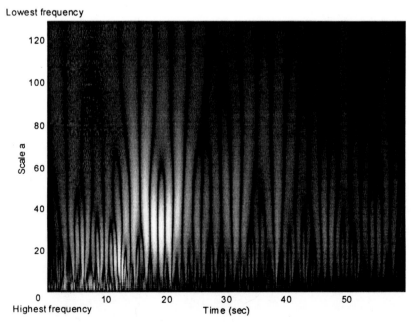

Figure 4.17 Absolute values of the wavelet coefficients for the 1971 San Fernando earthquake (scales = 1 to 128)

Chapter 5
Feedback Control Algorithms

5. 1. Introduction

Since Yao (1972) introduced the control concept to structural engineers, a number of control algorithms have been used for structural problems. The linear quadratic regulator (LQR) feedback control algorithm (Soong, 1990) and the linear quadratic Gaussian (LQG) control algorithm (Stein and Athans, 1987; Dyke et al., 1996b) are among the most popular optimal feedback control algorithms mainly due to their simplicity and ease of implementation. These algorithms achieve a significant level of attenuation in the vicinity of the natural frequencies of the structure. However, they fail to suppress the vibrations when the frequency of the external disturbance differs from the natural frequencies of the structure. Further, these algorithms are susceptible to parameter uncertainty and modeling error (Prakah-Asante and Craig, 1994) and they present optimum solutions in a narrow sense only because the external excitation term is ignored in their formulation and solution. In these algorithms, a pre-defined performance index is minimized where only the responses of the system and control effort are included.

Yang et al. (1987) attempted to include the external excitation in the formulation by proposing an instantaneous optimal control algorithm that minimizes the performance index at every instant of time within the control interval. This algorithm, however, is much more dependent on the choice of weighting matrices than the LQR and LQG algorithms, thus requiring careful consideration in order to achieve desirable control results (Soong, 1990).

Further, the stability of the control system is not guaranteed (Yang and Li, 1991). Suhardjo et al. (1992) include the external excitation in the frequency domain in an optimal feedback-feedforward control algorithm in their study of control of wind-excited building structures. Wind loads are modeled in the frequency domain as stochastic processes by their spectral density matrices. The authors combine the external wind loads with a feedback controller in the form of feedforward filters in the formulation of the control problem.

To overcome the aforementioned limitations of the classic optimal control algorithms, researchers have explored the use of soft computing approaches such as neural networks and fuzzy logic (Adeli and Hung, 1995). Neural networks are capable of learning and generalizing (Adeli and Park, 1998; Adeli and Karim, 2001; Adeli and Yeh, 1989; Hung and Adeli, 1991a and b; Adeli and Zhang, 1993; Adeli and Hung, 1993a and b; Hung and Adeli, 1993; Adeli and Hung, 1994; Hung and Adeli, 1994; Adeli and Park, 1995a and b). A review of the civil engineering applications of neural networks is presented by Adeli (2001). The backpropagation (BP) neural network learning algorithm is the most widely used neural network algorithm because of its simplicity. Ghaboussi and Joghataie (1995) and Chen et al. (1995) present active control algorithms using the BP neural networks. The BP algorithm is used first to predict the desired responses subjected to control forces and again to predict the control forces given the desired responses and the external excitation. Chen et al. (1995) define the instantaneous error function as the summation of error between actual and desired responses. Then, the BP training rule is applied to minimize the error function. The desired response is set to zero in each time step. Ghaboussi and Joghataie (1995), however, set the average of expected responses for a few future time steps to zero. As such, in the BP-based control algorithm, the desired output is selected somewhat arbitrarily and may not be optimal. The

BP algorithm is used for function approximation. To achieve satisfactory results, a hidden layer with a large number of nodes is needed resulting in a very slow learning process and a very large number of iterations for solution convergence. Moreover, the BP algorithm suffers from the hill climbing problem, that is, the solution can be trapped in a local minimum during the training (Bakshi and Stephanopoulos, 1993; Adeli and Hung, 1994).

5. 2. Equation of motion

A major reason for the use of active control is to minimize the displacements and stresses under severe dynamic loading conditions. As such, the structural response will be limited to the elastic range. The control algorithms presented in this section are based on the assumption that the structure is time-invariant and behaves linearly.

When an m-degree-of-freedom (DOF) discrete system is subjected to external excitation and control forces, its governing equation of motion can be written as (Soong, 1990)

$$M\ddot{u}(t) + C\dot{u}(t) + Ku(t) = B_c f(t) + E_c f_e(t) \tag{5.1}$$

where $f(t) = l \times 1$ control force vector; B_c and E_c are $m \times l$ and $m \times r$ location matrices which define locations of the control forces and the external excitations, respectively, and t is the time. In state-space form, Eq. (5.1) can be written in the form

$$\dot{z}(t) = Az(t) + Bf(t) + Ef_e(t) \tag{5.2}$$

where

$$z(t) = \begin{bmatrix} u(t) \\ \dot{u}(t) \end{bmatrix} \tag{5.3}$$

is the $2m \times 1$ state vector, and

$$A = \begin{bmatrix} 0 & I \\ -M^{-1}K & -M^{-1}C \end{bmatrix} \tag{5.4}$$

$$B = \begin{bmatrix} 0 \\ M^{-1}B_c \end{bmatrix} \tag{5.5}$$

$$E = \begin{bmatrix} 0 \\ M^{-1}E_c \end{bmatrix} \tag{5.6}$$

are $2m \times 2m$, $2m \times l$, and $2m \times r$ system, control location, and external excitation location matrices, respectively. The matrices 0 and I in Eqs. (5.4) to (5.6) denote, respectively, the zero and identity matrices of size $m \times m$.

5. 3. LQR control algorithm

The LQR optimal control algorithm is one of the most widely used feedback algorithms in structural control mainly due to its simplicity and relative ease of implementation (Adeli and Saleh, 1999; Kurata et al., 1999). The optimal control is defined by a given vector of controllers and predefined state variable performance weighting matrix, Q, and control effort weighting matrix, R. The problem is then expressed as finding the appropriate state-feedback control forces that minimize the following performance index:

$$J = \int_0^\infty \left[z^T(t)Qz(t) + f^T(t)Rf(t) \right] dt \tag{5.7}$$

where the superscript ^{T} denotes the transpose of a matrix. Then, optimal state-feedback control forces are obtained from

$$f(t) = -Gz(t) = -R^{-1}B^T Pz(t) \tag{5.8}$$

where G is the gain matrix and P is obtained from the solution of the algebraic Riccati equation:

$$- PA - A^T P - Q + PBR^{-1} B^T P = 0 \qquad (5.9)$$

where the superscript $^{-1}$ denotes the inverse of a matrix. The solution of the Riccati equation can be obtained by the generalized eigenproblem algorithm (Arnold, 1984) or other methods (Lewis and Syrmos, 1995; Saleh and Adeli, 1997). Substituting Eq. (5.8) into Eq. (5.2), the behavior of the optimally controlled structure can be obtained by

$$\dot{z}(t) = (A - BG)z(t) + Ef_e(t) \qquad (5.10)$$

5.3.1. Application to active tuned mass damper

The example structure considered in this chapter is the active tuned mass damper (ATMD) control model, shown in Figure 5.1, presented at the web site "Java Powered Simulator for Structural Vibration and Control" (Yang and Satoh, 2001). Structural properties of the ATMD system and the weight coefficients are

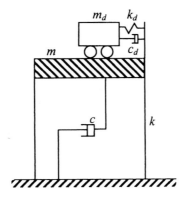

Figure 5.1 Active tuned mass damper (ATMD) system

$$M = \begin{bmatrix} 100 & 0 \\ 0 & 1 \end{bmatrix} \times 10^3 \, kg \tag{5.11}$$

$$C = \begin{bmatrix} 1.3495 & -0.0928 \\ -0.0928 & 0.0928 \end{bmatrix} \times 10^4 \, \text{N} \cdot \text{s/m} \tag{5.12}$$

$$K = \begin{bmatrix} 3.9861 & -0.0383 \\ -0.0383 & 0.0383 \end{bmatrix} \times 10^6 \, \text{N/m} \tag{5.13}$$

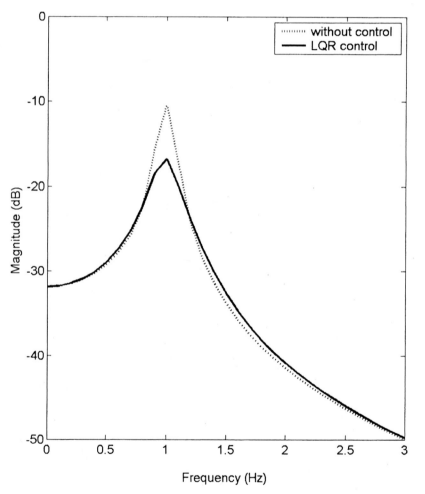

Figure 5.2. Frequency responses of the ATMD system with and without LQR control

$$Q = diag[5000 \quad 10 \quad 0 \quad 0] \tag{5.14}$$

$$R = [1] \tag{5.15}$$

The fundamental natural frequency, ω_n, of the main system is 2π rad/sec, i.e., $f_n = 1$ Hz. Figure 5.2 shows the comparison of the frequency responses of the ATMD system with and without LQR control in decibels (dB). Decibel is a logarithmic unit defined as 20log10X, where X is the root

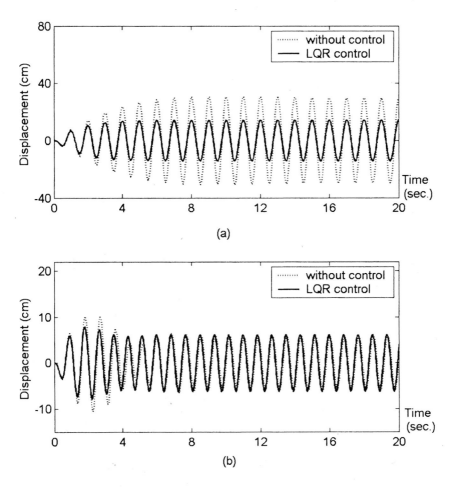

Figure 5.3 LQR control of the ATMD system: (a) $\omega = 1.0\omega_n$, (b) $\omega = 1.2\omega_n$

mean square quantity. The gain matrix G calculation is done using the Matlab (2000) LQR routine. As seen in Figure 5.2, a significant level of attenuation is achieved in the vicinity of the resonance frequency of the ATMD system. However, the level of attenuation reduces drastically when the frequency of the external disturbance differs from the fundamental frequency of the ATMD system.

Consequently, the LQR control method results in very little suppression of vibration for the latter case, as demonstrated in Figure 5.3. Figure 5.3(a) shows the displacement of the system when the disturbance frequency, ω is same as the natural frequency of the system and substantial suppression of vibrations is achieved. In contrast, Figure 5.3(b) shows the displacement when the disturbance frequency is 1.2 times the fundamental frequency of the ATMD system. In this case, the vibration suppression is minimal and diminishes as time goes on.

Table 5.1 Responses of the ATMD system subjected to disturbances with frequencies the same as and 1.2 times the natural frequency of the system

Disturb-ance frequency	Uncontrolled responses		LQR controlled responses	
	Maximum displacement (cm)	RMS displacement (cm)	Maximum displacement (cm)	RMS displacement (cm)
$\omega = 1.0\ \omega_n$	30.9	20.0	14.4 (53.2%)	9.84 (50.8%)
$\omega = 1.2\ \omega_n$	10.7	4.43	8.02 (24.9%)	4.46 (-0.68 %)

Table 5.1 summarizes the maximum responses and RMS displacements with reduction ratios presented in parentheses. The results clearly show that the control effectiveness decreases considerably when the

frequency of the external disturbance differs from the fundamental frequency of the ATMD system.

5. 4. LQG control algorithm

Another commonly used feedback control algorithm in structural control is the LQG control algorithm (Dyke et al., 1996a; Spencer et al., 1998). In this approach, the measured outputs are assumed to be the desired system response plus noise. This consideration is due to the fact that there are inherent errors in the structure modeling as well as in the output sensing. Considering noise in the measured response, the controlled response, y_c, and measured response, y_m, are given by

$$y_c = C_c z + D_c f + F_c f_e \tag{5.16}$$

and

$$y_m = C_m z + D_m f + F_m f_e + v \tag{5.17}$$

respectively, where C_c, D_c, F_c, C_m, D_m, and F_m are mapping matrices with appropriate dimensions and v is the measurement noise vector.

For the LQG feedback control algorithm, the optimal control problem is expressed as finding the appropriate state-feedback control forces that minimize the following performance index:

$$J = E\left\{ \lim_{t_f \to \infty} \frac{1}{t_f} \int_0^{t_f} \left[(C_c z + D_c z)^T Q (C_c z + D_c z) + f^T R f \right] dt \right\} \tag{5.18}$$

where $E\{\}$ denotes the expected value operator. The control force is obtained as

$$f = -G\hat{z} \tag{5.19}$$

in which $\hat{z}$ is the Kalman filter estimator of the state vector, which is given by

$$\dot{\hat{z}} = A\hat{z} + Bf + L\left(y_m - C_m\hat{z} - D_m f\right) \qquad (5.20)$$

where matrix L is determined by using the standard Kalman filter estimator technique (Dorato et al., 1995; Spencer et al., 1998; Skelton, 1988). Substituting Eq. (5.19) into Eq. (5.20) yields the closed-loop form as

$$\dot{\hat{z}} = \left(A - BG - LC_m + LD_m G\right)\hat{z} + Ly_m \qquad (5.21)$$

The frequency response of the ATMD system using the LQG control is not presented here because it is very similar to that shown in Figure 5.2 for the LQR method. The LQG control method also suppresses the vibrations effectively only when the external disturbance frequency is near the fundamental frequency of the system.

5. 5. Shortcomings of classic control algorithms

Both the LQR and LQG control algorithms are sensitive to structural modeling and discretization errors and vibrations in the sensoring equipment (Prakah-Asante and Craig, 1994). They present optimum solutions in a narrow sense only because the external excitation term is ignored in their formulation and solution. In these algorithms, a pre-defined performance index is minimized where only the responses of the system and control effort are included. This limitation of classical optimal control algorithms is due to the fact that the input excitation must be known *a priori* which is not the case for earthquake or wind loads.

Chapter 6
Filtered-x LMS Algorithm

6. 1. Introduction

The adaptive filtered-x least mean square (LMS) control algorithm has been used successfully in acoustic, electrical, and aerospace engineering problems (Widrow and Stearns, 1985). This algorithm is based on the integration of the adaptive filter theory used for system identification in real time and the feedforward control approach. The advantage of this method is that the external excitation is included in the formulation. This algorithm was used by Burdisso et al. (1994) for active control of a three-story two-dimensional frame subjected to earthquake loading. They point out that this algorithm can handle the modeling error including the effect of soil-structure interaction. Since the control forces determined are adapted by updating the finite-impulse-response (FIR) filter coefficients at each sampling time until the output error is minimized, the filtered-x LMS control scheme minimizes vibrations over the entire frequency range and thus is less susceptible to modeling errors and inherently more stable. However, it is not as effective for short transient vibrations such as peaks because it requires adaptation time.

6. 2. Adaptive LMS filter

The adaptive LMS filter algorithm was developed in the system identification field (Widrow and Stearns, 1985). Figure 6.1 shows an adaptive filter in the form of system identification. An external input signal,

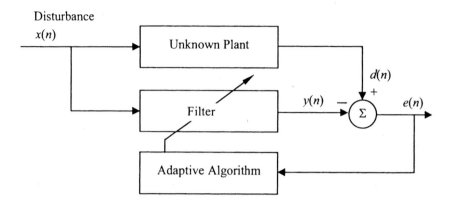

Figure 6.1 Adaptive filter adjusted to emulate the response of an unknown system

$x(n)$, is fed into both the unknown system and the filter, and the outputs of the unknown system and filter, $d(n)$ and $y(n)$, are subtracted to find an error signal, $e(n)$:

$$e(n) = d(n) - y(n) \qquad (6.1)$$

where n is an integer defining the nth discrete time step. The aim of the adaptive algorithm is to adapt the filter coefficients such that the error sequence is as close to zero as possible in a squared mean sense.

When the FIR filter is used, the output of an FIR filter is expressed in terms of input as

$$y(n) = \sum_{i=0}^{L-1} w_i(n)x(n-i) = w(n)^T x(n) \qquad (6.2)$$

where L = order of filter; w_i = ith coefficient; $w(n) = [w_0(n)\ w_1(n)\ \ldots\ w_{L-1}(n)]^T$ = coefficient vector; and $x(n) = [x(n)\ x(n-1)\ \ldots\ x(n-L+1)]^T$ = input signal vector.

In the adaptive LMS filter, the coefficients vector $w(n)$ is adapted by using the LMS algorithm to minimize the error signal, $e(n)$. A cost function to be minimized is defined by

$$J(n) = E\{e(n)\}^2$$

$$= E\{d(n) - w^{\mathrm{T}}(n)x(n)\}^2$$

$$= E\{d(n)^2\} - 2w^{\mathrm{T}}(n)R_{dx} + w^{\mathrm{T}}(n)R_{xx}w(n) \qquad (6.3)$$

where

$$R_{dx} = E\{d(n)x(n)\} \qquad (6.4)$$

$$R_{xx} = E\{x(n)x(n)^{\mathrm{T}}\} \qquad (6.5)$$

The square matrix R_{xx} is the input correlation matrix, and the vector R_{dx} is the set of cross-correlation between the desired response and the input signals.

Widrow and Stearns (1985) proposed the simple and effective LMS algorithm to find the minimum mean-squared error as

$$w_i(n+1) = w_i(n) - \mu_G \frac{\partial J(n)}{\partial w_i(n)} \qquad (6.6)$$

where μ_G = gain constant that regulates the speed and stability of the adaptive algorithm. Taking derivatives of $J(n)$ with respect to the elements of $w(n)$ yields

$$\frac{\partial J(n)}{\partial w_i(n)} = 2E\left\{e(n)\frac{\partial e(n)}{\partial w_i(n)}\right\} \qquad i = 0,\ldots,L\text{-}1 \qquad (6.7)$$

Using the instantaneous values to approximate expected values of the gradient, Eq. (6.7) can be simplified as

$$\frac{\partial J(n)}{\partial w_i(n)} = 2e(n)\frac{\partial e(n)}{\partial w_i(n)} \qquad i = 0,\ldots,L\text{-}1 \qquad (6.8)$$

Combining Eqs. (6.1), (6.2) and (6.8) we obtain

$$\frac{\partial J(n)}{\partial w_i(n)} = -2e(n)x(n-i) \qquad i = 0,\ldots,L\text{-}1 \qquad (6.9)$$

Substituting Eq. (6.9) into Eq. (6.6) we find

$$w_i(n+1) = w_i(n) + 2\mu_G e(n)x(n-i) \quad i = 0,\ldots,L\text{-}1 \qquad (6.10)$$

Or, in vector form

$$w(n+1) = w(n) + 2\mu_G e(n)x(n) \qquad (6.11)$$

The bound on μ_G for stability is derived as (Widrow and Stearns, 1985)

$$0 < \mu_G < \frac{1}{\lambda_{\text{max}}} \qquad (6.12)$$

where λ_{max} is the largest eigenvalue of the input correlation matrix R_{xx}. In this work, we use the normalized-LMS (NLMS) algorithm where the constant μ_G is substituted by a time-varying function $\mu_G(n)$ defined as

$$\mu_G(n) = \frac{1}{a + x(n)^{\text{T}} x(n)} \qquad (6.13)$$

in which a = a small positive constant to overcome the potential numerical instability in the filter coefficients update. The advantage of using the NLMS algorithm over the constant LMS algorithm is that the adaptation is inherently stable (Tarrab and Feuer, 1988).

Figure 6.2 summarizes the above process by showing an FIR filter that is updated using the NLMS algorithm. Figure 6.3 shows the simulation of the system identification process using a single-frequency infinite harmonic disturbance with a sampling frequency of 50 Hz (time step of 0.02 seconds)

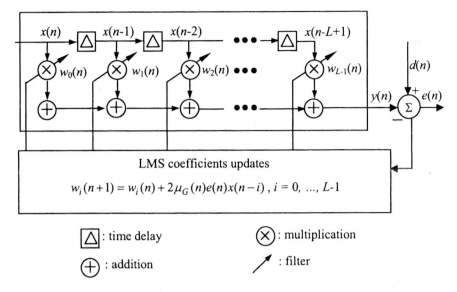

Figure 6.2 Adaptive FIR filter updated using the NLMS algorithm

for the ATMD system presented in section 5.3.1. Figure 6.3(a) shows the simulated response of the ATMD system to the single-frequency harmonic motion. Figure 6.3(b) shows the error signal as defined by Eq. (6.1). It is observed that the error signal is reduced to zero after about 20 seconds (that means the system is fully identified after 20 seconds). Adaptation of two filter coefficients (w_i) over time is presented in Figure 6.4. Note that a second order filter with two filter coefficients is sufficient for identification of the ATMD system subjected to a single-frequency sinusoidal external disturbance.

6. 3. Filtered-x LMS control algorithm

The direct form of the LMS algorithm cannot be used for active control of structures because the response of the system depends not only on the external disturbance but also on the control forces. In Figure 6.5 the entire plant is divided into two systems: the structural and control systems. The structural system represents the external disturbance-to-output relationship of the plant expressed with state-space matrices A and E and the control system represents the control force-to-output relationship of the plant expressed with state-space matrices A and B in Eq. (6.2). In practice, the control force-to-output relationship is estimated by an FIR or infinite-impulse-response (IIR) filter coefficient, and these filter coefficients are obtained in the offline LMS implementation. In the following paragraphs, for the derivation of the

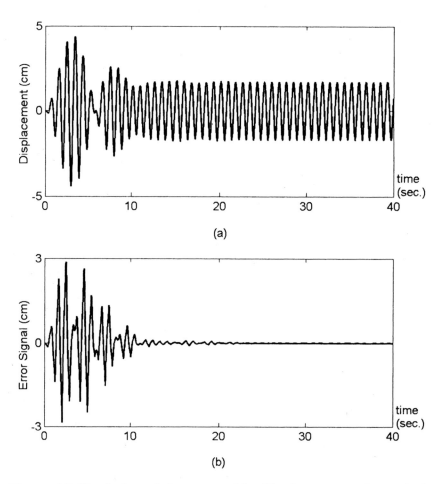

Figure 6.3 Simulation of the system identification process for a single harmonic disturbance on the ATMD system: (a) displacement, $d(n)$, (b) error signal, $e(n)$

filtered-x LMS control algorithm, this relationship denoted by discrete filter coefficients $h_c(n)$ in Figure 6.5 is estimated by an FIR filter of order K.

The output of the control filter $f_x(n)$ (the input to the control system) is expressed in terms of the FIR filter coefficients $w_i(n)$ and the external disturbance input signal $x(n-i)$ as

$$f_x(n) = \sum_{i=0}^{L-1} w_i(n)x(n-i) \tag{6.14}$$

The output of the control system due to control input only is obtained as follows (Widrow and Stearns, 1985):

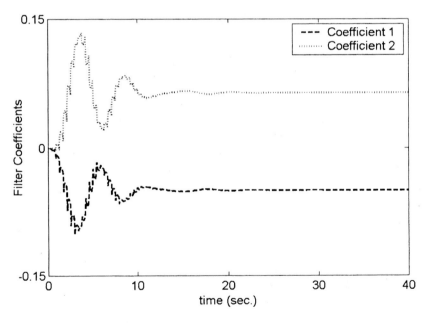

Figure 6.4 Adaptation of two filter coefficients during simulation

$$y_c(n) = \sum_{j=0}^{K-1} h_c(j) f_x(n-j)$$

$$= \sum_{j=0}^{K-1} h_c(j) \sum_{i=0}^{L-1} w_i(n) x(n-i-j) \tag{6.15}$$

Then, the output error, the net output of the system $e(n)$, is taken as the difference between the output of the structural system $y_s(n)$ and that of the control plant $y_c(n)$:

$$e(n) = y_s(n) - \sum_{j=0}^{K-1} h_c(j) \sum_{i=0}^{L-1} w_i(n) x(n-i-j) \tag{6.16}$$

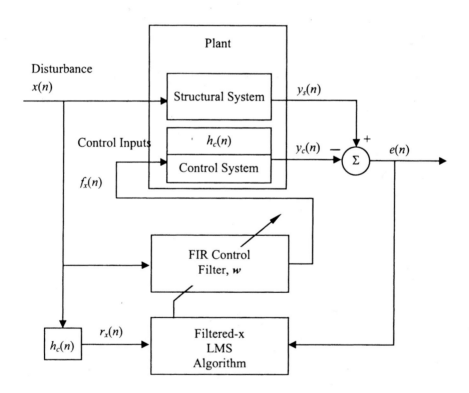

Figure 6.5 Adaptive filter being updated using the filtered-x LMS algorithm

Since the order of convolution can be interchanged without affecting the summation result, we can rewrite Eq. (6.16) as

$$e(n) = y_s(n) - \sum_{i=0}^{L-1} w_i(n) \sum_{j=0}^{K-1} h_c(j) x(n-i-j) \qquad (6.17)$$

which can be further simplified as

$$e(n) = y_s(n) - \sum_{i=0}^{L-1} w_i(n) r_x(n-i) \qquad (6.18)$$

where

$$r_x(n-i) = \sum_{j=0}^{K-1} h_c(j) x(n-i-j) \qquad (6.19)$$

This procedure to produce the resultant output r_x by rearranging convolution is dubbed the filtered-x operation (Widrow and Stearns, 1985). The term "filtered-x" refers to the input signal x.

The gradient of the cost function $J(n)$ defined by Eq. (6.3) then becomes

$$\frac{\partial J(n)}{\partial w_i(n)} = -2r_x(n-i)e(n) \qquad i = 0,\ldots, L\text{-}1 \qquad (6.20)$$

The filter coefficients of the control system are updated and adapted according to Eq. (6.6) yielding

$$w_i(n+1) = w_i(n) + 2\mu_G(n) r_x(n-i)e(n) \quad i = 0,\ldots, L\text{-}1 \qquad (6.21)$$

The control forces are adapted at each sampling time using Eq. (6.19) with the updated filter coefficients obtained from Eq. (6.21) until the output error is minimized. In other words, the filtered-x LMS algorithm finds an optimal value of the cost function in real time by adapting its values of coefficients, while the cost function (performance index) of the LQR/LQG control algorithm is optimized offline.

6. 4. Application to active tuned mass damper

Figure 6.6 presents the results of filtered-x LMS control for the same ATMD system. The frequency of the external disturbance is set to 1.2 times the

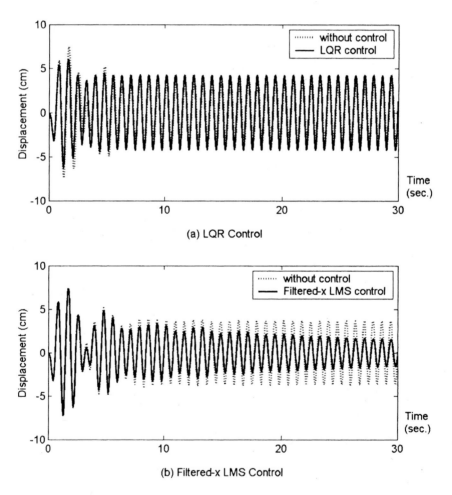

(a) LQR Control

(b) Filtered-x LMS Control

Figure 6.6 Filtered-x LMS control of the ATMD system, $\omega = 1.2\omega_n$

fundamental frequency of the system ($\omega = 1.2\omega_n$) where both LQR and LQG control algorithms are limited in suppressing the vibration, as discussed in sections 5.3 and 5.5. Figure 6.6(a) is the same as Figure 5.3(b) and is presented here again for better comparison. At the beginning, as seen in Figure 6.6(b), the filtered-x LMS control algorithm shows little vibration suppression compared to the LQR control algorithm, but more and more vibration suppression is achieved as time goes on. This is due to the fact that the filtered-x LMS control algorithm requires time of filter adaptation for control. Consequently, it can be concluded that this filtered-x LMS algorithm is not as effective for short transient vibrations such as peaks since it requires adaptation time, but is effective in suppressing system vibration outside the resonance frequency.

Chapter 7

Hybrid Feedback-LMS

Algorithm

7. 1. Introduction

A hybrid feedback-LMS control algorithm is introduced in this chapter combining the feedback and filtered-x LMS control algorithms. The feedback control methods are susceptible to modeling errors, which affect their stability, as described earlier. Though the filtered-x LMS control scheme minimizes vibrations over the entire frequency range and thus is less susceptible to modeling errors and inherently more stable, it is not as effective for short transient vibrations such as peaks since it requires adaptation time.

7. 2. Hybrid feedback-LMS algorithm

The hybrid feedback-LMS control algorithm introduced in this chapter is intended to minimize vibrations for both steady state and transient vibrations by combining the feedback control together with a robust adaptive filtered-x LMS algorithm. The resulting new algorithm is robust because it takes into account different external disturbances and a large frequency range. The hybrid control algorithm, shown in Figure 7.1, can be a combination of LQR or LQG and the filtered-x LMS algorithms. In this algorithm, the external disturbance signal, $x(n)$, is simultaneously fed into the structural system and filtered-x LMS adaptive controller. The control force, $f_x(n)$, obtained through

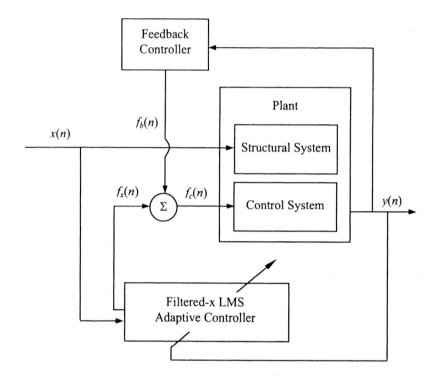

Figure 7.1 Hybrid feedback-LMS control

the filtered-x LMS adaptive controller is added to the feedback control force, $f_b(n)$, to yield the total control force, $f_c(n)$, and applied to the structural system to be controlled. The response of the structure, $y(n)$, is then fed back into both the feedback controller to obtain the feedback control force, $f_b(n)$, and the filtered-x LMS adaptive controller to update FIR filters and obtain the control force, $f_x(n)$.

7. 3. Application to active tuned mass damper

Shown in Figure 7.2(a) is the response of the ATMD system subjected to an external disturbance with a frequency equal to 1.2 times the fundamental frequency of the system using the hybrid feedback-LMS control algorithm.

The FIR filter of order 10 is used in the control model. In this example, a full-state feedback LQR controller is combined with a filtered-x LMS algorithm where the displacement of the main structure is used as the error signal in Eq. (6.15). The algorithm can be easily modified for velocity- or acceleration-feedback control for more realistic applications. This is demonstrated in the next section where a few selected acceleration responses are used as feedback states for the LQG controller as well as error signals to the filtered-x LMS adaptive controller.

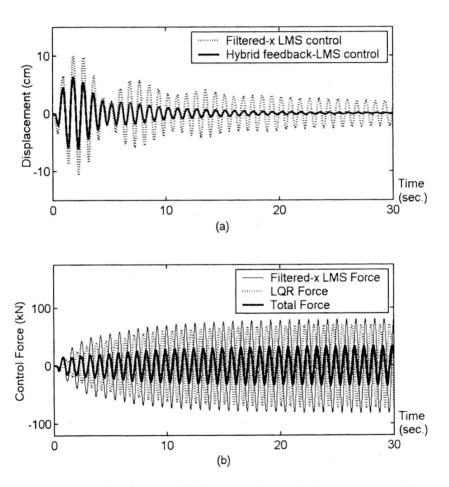

Figure 7.2 Hybrid feedback-LMS control of the ATMD system, $\omega = 1.2\,\omega_n$

The dotted line in Figure 7.2(a) is the results for the filtered-x LMS control algorithm [the same as the solid line in Figure 6.6(b)] and the solid line is the results for the hybrid algorithm. It is observed that the hybrid feedback-LMS control algorithm achieves faster vibration suppression than the filtered-x LMS algorithm. Moreover, responses in earlier stages are similar to those shown in Figure 6.6(a), that is, transient vibrations are controlled by the LQR controller and thus more vibration suppressions than the filtered-x LMS algorithm are made. The same conclusion is made from Table 7.1 where response results of the ATMD system for different control algorithms are summarized with reduction ratios presented in parentheses when the disturbance frequency is 1.2 times the fundamental frequency of the ATMD system. While the LQR and filtered-x LMS algorithms can effectively reduce either maximum displacement or RMS displacement, respectively, the hybrid feedback-LMS control algorithm can achieve significant reductions in both maximum and RMS displacements.

The control force for the hybrid feedback-LMS control is presented in Figure 7.2(b). The total control force, $f_c(n)$, is sum of the filtered-x LMS force, $f_x(n)$, plus the LQR force, $f_b(n)$. We can see that the envelop of the total control force increases until the displacement of the ATMD system approaches zero, that is, until the filter coefficient updates are stabilized.

Table 7.1 Responses of the ATMD system using $\omega = 1.2\ \omega_n$

Control algorithm	Maximum displacement (cm)	RMS displacement (cm)
No control	10.68	4.46
LQR	8.02 (24.9%)	4.46 (-0.68%)
Filtered-x LMS	10.4 (2.53%)	3.04 (31.4%)
Hybrid feedback-LMS	6.36 (40.5%)	1.38 (69.0%)

7. 4. Concluding remarks

Figures 7.3(a) to 7.3(d) show the responses of the ATMD system subjected to an external disturbance with a frequency equal to 1.5 times the fundamental frequency of the system without control, and with LQR, filtered-x LMS, and the new hybrid control algorithms, respectively. These figures show while the LQR control algorithm results in a slight increase of the steady state response, the new control algorithm results in consistent vibration suppression in both transient and steady state responses. This is due to the fact that the external disturbance is a sinusoidal signal with only one frequency component and therefore updating of filter coefficients is not affected by the frequency of the external disturbance by any significant measure

Figure 7.4 shows a comparison of different orders of FIR filters for the ATMD system identification subjected to white noise. White noise has an infinite number of frequencies thus requiring an infinite number of filter coefficients for exact identification. For FIR filters with a smaller number of filter coefficients, larger ripples (errors) are created when the frequency differs more from the natural frequency of the ATMD system.

When the new hybrid feedback-LMS control algorithm is used to control realistic structures against actual destructive environmental forces such as earthquake loads, the adaptation of filter coefficients takes much longer and the required number of filter coefficients becomes large. This is due to the fact that the environmental loads are wideband signals. Moreover, the accurate estimation of the properties of actual structures with the FIR filter involves a large number of filter coefficients.

104

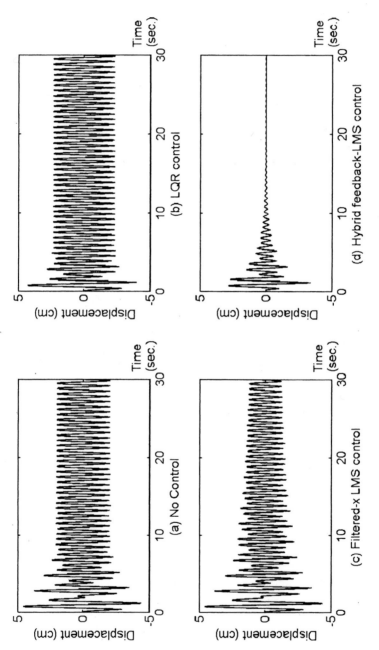

Figure 7.3 Responses of the ATMD system subjected to an external disturbance with a frequency equal to 1.5 times the fundamental frequency of the system

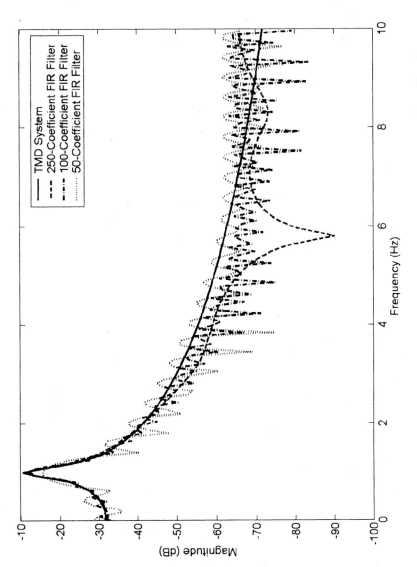

Figure 7.4 Comparison of different coefficient FIR filters for system identification

As observed in Figure 7.4 even for a simple system, a large number of filter coefficients is required to achieve a close approximation to the actual frequency response of the system, requiring a significant amount of computational time. Also, note in Figure 7.4 that FIR filters with larger numbers of filter coefficients show poorer approximations in the higher range of frequencies. In the next chapter, the hybrid feedback-LMS algorithm presented in this chapter is extended for control of structures subjected to realistic environmental forces through integration with a discrete wavelet low-pass filter.

Chapter 8

Wavelet-Hybrid Feedback-LMS Algorithm for Robust Control of Structures

8. 1. Introduction

An advantage of the adaptive filtered-x LMS control algorithm is that the external excitation is included in the formulation. Further, it can successfully suppress vibrations due to an external disturbance whose frequency differs from the natural frequencies of the structure as demonstrated in Chapter 5. Though the filtered-x LMS control scheme minimizes vibrations over the entire frequency range and thus is less susceptible to modeling errors and inherently more stable, it is not as effective for short transient vibrations such as peaks because it requires adaptation time.

The hybrid feedback-LMS control algorithm, introduced in Chapter 6, is intended to minimize both steady state and transient vibrations by combining the feedback control with a robust adaptive filtered-x LMS algorithm. It is shown that the hybrid feedback-LMS control algorithm achieves faster vibration suppression than the filtered-x LMS algorithm. Further, the algorithm is robust because it takes into account different external disturbances and a large frequency range.

The hybrid feedback-LMS control algorithm, however, cannot be applied directly for control of realistic structures against actual destructive environmental forces such as wind, ocean waves, and earthquake loads. The frequency bandwidths of those environmental forces are much wider than the frequency bandwidth of common structural systems. Moreover, as noted in Chapter 6, in the FIR filter used in the filtered-x LMS algorithm, a large number of filter coefficients are required to achieve a close approximation to the actual frequency response of the system, requiring a significant amount of computational time and making the real time control of structures impractical. Also, FIR filters with larger numbers of filter coefficients show poorer approximations in the higher range of frequencies.

In this chapter, this shortcoming is overcome by integrating a low-pass filter with the filtered-x LMS adaptive controller. A low-pass filter passes all lower frequency signal components with frequencies from zero to the filter cutoff frequency unchanged. Higher frequency components above that cutoff frequency are eliminated, reducing the signal disturbance. Keeping signal components only within certain frequency limits helps the hybrid feedback-LMS control algorithm adapt its coefficients in a more stable fashion by eliminating higher frequency components that obstruct the stabilization of coefficients. This can be effective because the response of most civil structures is not affected by high frequency contents of the external excitations by any significant measure (the exception can be very rigid structures).

Low-pass filtering of signals commonly is made in the Fourier domain. Wang and Wu (1995) use a fourth-order Butterworth dual channel low-pass filter for structural system identification using the LMS method. This filter is often used for low frequency digital signal processing applications, where the sampling frequency (inverse of the discrete time step size) is much higher than the data bandwidth. Though low-pass filters in the Fourier domain are suitable for system identification where real time implementation is not an issue, they are not appropriate for real time control of structures because of their inordinate computational requirements.

In this chapter, a wavelet based low-pass filtering is presented. Considering the fact that the orthogonal wavelet filtering requires only integer operations, real time control of large structures can be achieved with little additional computational efforts due to filtering.

8. 2. Wavelet transforms as an effective filter for control problems

Based the concept of the filter bank shown in Figure 3.3 and Eqs. (3.12) and (3.13), wavelet transform can be used for low-pass or high-pass filtering depending on the application. An example of wavelet filtering using a Daubechies wavelet with 3 vanishing moments is illustrated in Figures 8.1 and 8.2. The original signal shown in Figure 8.1(a) is a hypothetical ground acceleration signal composed of a sinusoidal signal with a frequency of 1.2 Hz and white noise with a standard deviation of 0.2. The sampling frequency is 50 Hz. The original signal in Figure 8.1(a) is low-passed and high-passed up to the second level of the filter bank (Figure 3.3). Figures 8.1(b) and 8.1(c) show high-pass and low-pass filtered signals of the original signal shown in Figure 8.1(a). The low-pass filtered signal shown in Figure 8.1(c) approximates the original sinusoidal signal because low-pass filtering reduces the noise in this particular application. The signal in Figure 8.1(b) is the difference between the signals in Figures 8.1(a) and 8.1(c) and represents the eliminated noise.

The Fourier transforms of the hypothetical ground acceleration signal shown in Figure 8.1(a) as well as the high-pass and low-pass filtered signals shown in Figures 8.1(b) and 8.1(c) are presented in Figures 8.2(a) to 8.2(c), respectively. As expected, the peak amplitude in Figure 8.2(a) occurs at a frequency of 1.2 Hz, the dominant frequency of the original signal shown in

Figure 8.2(a). Figure 8.2(b) shows that signals with frequencies roughly below 5 Hz are filtered by the high-pass filtering. Figure 8.2(c) shows that signals with frequencies roughly above 7 Hz are filtered by the low-pass filtering. For an ideal filter, the cutoff frequency for both high- and low-pass filtering should be the same. As for most practical filters, however, the cutoff frequencies of the high-pass and low-pass wavelet filters do not coincide. The difference between the cutoff frequencies depends on the type of wavelet used.

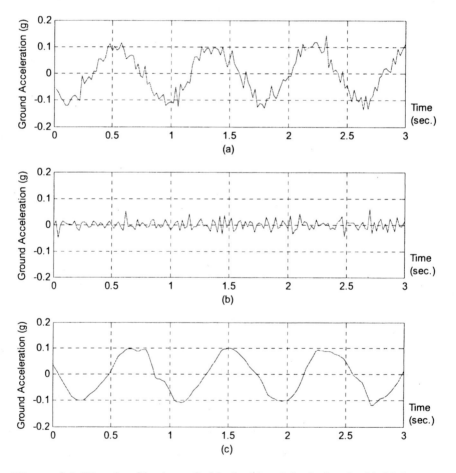

Figure 8.1 Wavelet filtering of signal: (a) original signal, (b) high-pass filtered signal, (c) low-pass filtered signal

The need for low-pass filtering for effective control of civil structures subjected to extreme environmental forces was discussed earlier. The examples presented in this section and results displayed in Figures 8.1 and 8.2 indicate that the wavelet transform can be used as an effective filtering scheme for structural control problems.

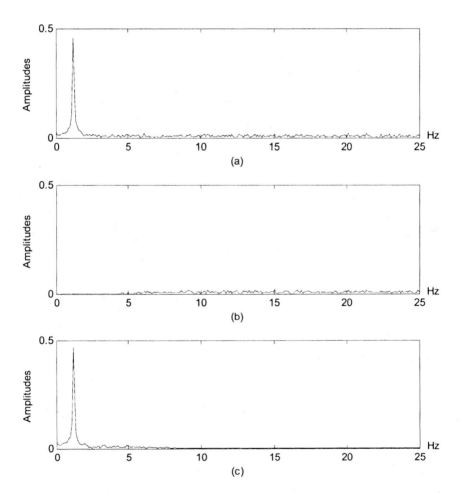

Figure 8.2 Fourier transforms of signals shown in Figure 8.1: (a) original signal, (b) high-pass filtered signal, (c) low-pass filtered signal

112

8. 3. Wavelet-hybrid feedback-LMS control algorithm

Figure 8.3 shows the architecture of the proposed wavelet-hybrid feedback-LMS control algorithm. The external disturbance signal, $x(n)$, is simultaneously fed into the structural system without filtering and into the filtered-x LMS adaptive controller after being filtered by the wavelet low-pass filter. The wavelet low-pass filtered signal is fed into the filtered-x LMS adaptive controller to obtain the control force $f_x(n)$. This force is then added to the feedback control force, $f_b(n)$, to yield the total control force, $f_c(n)$, and applied to the structural system to be controlled. The response of the structure is then fed back into both the feedback controller to obtain the feedback control force, $f_b(n)$, and the filtered-x LMS adaptive controller to update the FIR filter coefficients and obtain the control force, $f_x(n)$. It should

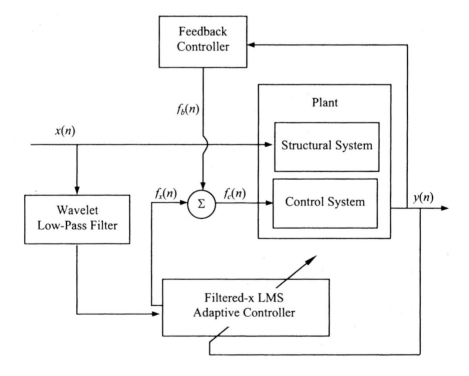

Figure 8.3 Architecture of the wavelet-hybrid feedback-LMS control model

be noted that in the model presented in Figure 8.3 the wavelet filtering affects only the filtered-x LMS adaptive controller and not the feedback controller. This is because the input to the feedback controller needs to be the response of the structural system subjected to unfiltered signals.

The choice of the level of low-pass filtering (Figure 3.3) depends on the type of the structure (e.g., rigid versus flexible structures) and the sampling frequency. The level of wavelet low-pass filtering is chosen such that the cutoff frequency is greater than the largest significant natural frequency of the structure. In this study it is found that the cutoff frequency of 1.5 to 2 times the largest significant natural frequency produces the best control results. This is a somewhat large range for cutoff frequencies because these cutoff frequencies of the low-pass wavelet filters cannot be specified exactly as they depend on many factors such as sampling frequency, the type of wavelet chosen, and number of vanishing moments.

In the subsequent sections, the effectiveness of the proposed wavelet-hybrid feedback-LMS control model is demonstrated by application to two examples. The first example is the ATMD system described in section 5.3.1. The feedback control algorithm used in this example is the LQR algorithm. The second example is the active mass driver (AMD) benchmark problem solved by a number of different investigators (Spencer, et al., 1998). The feedback control algorithm used in this example is the LQG algorithm.

8.3.1. Application to active tuned mass damper

The wavelet-hybrid feedback-LMS control algorithm is applied to the ATMD system described in section 5.3.1 subjected to the hypothetical ground acceleration signal shown in Figure 8.1(a). For the simulation, the full-state feedback LQR controller is combined with the filtered-x LMS algorithm where the displacement of the main structure is used as the error

114

signal. Figures 8.4(a) to 8.4(c) show the uncontrolled response, the response with LQR control, and the response with the wavelet-hybrid feedback-LMS control model, respectively. The order of the FIR filter coefficients in the control model is set to 50, a relatively low number, to show the effectiveness

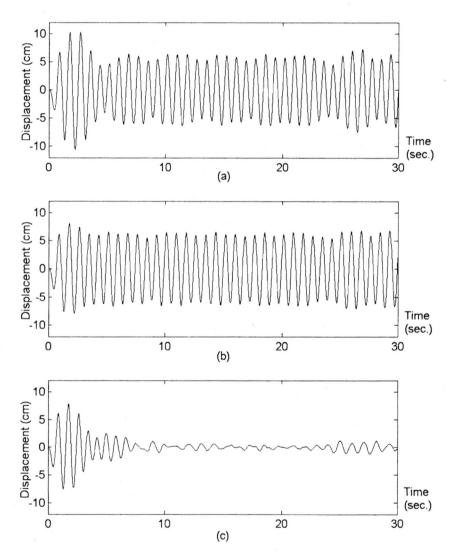

Figure 8.4 Response of the ATMD system: (a) uncontrolled response, (b) response with the LQR control, (c) response with the wavelet-hybrid feedback-LMS control

of the new model even though a 50-coefficient FIR filter does not provide a close approximation of the ATMD system as shown in Figure 7.4. Figure 8.4 demonstrates the superiority of the proposed model to classical LQR feedback control algorithms. It confirms that the proposed control model can effectively minimize the vibrations even when the bandwidth of the external disturbance (25 Hz) is much wider than the natural frequency of the structural system (1 Hz).

8.3.2. Application to an active mass driver (AMD) benchmark example structure

The proposed control model is applied to the AMD benchmark problem developed by the American Society of Civil Engineering (ASCE) Committee on Structural Control (Spencer et al., 1998; also see the web site at http://www.nd.edu/~quake). This structure is a two-dimensional scaled model of the prototype three-story building considered in Chung et al. (1989) and is subjected to two kinds of one-dimensional ground motion. An AMD is placed on the third floor of the structure to provide a control force to the structure. The AMD consists of a single hydraulic actuator with steel masses attached to the ends of the piston rod. The first three natural frequencies of the 3-story scaled frame are 5.81 Hz, 17.68 Hz, and 28.53 Hz.

Ten criteria, denoted by J_1 to J_{10}, are provided to evaluate the control performance. The first five performance measures (J_1 to J_5) are RMS responses of the structure and actuator subjected to an artificial ground acceleration record in the form of a stochastic signal with a spectral density defined by the Kanai-Tajimi spectrum. They are the RMS relative displacement of floors (J_1), the RMS acceleration of floors (J_2), and the RMS displacement, velocity, and acceleration of the actuator (J_3, J_4, and J_5, respectively). Three constraints are included: $\sigma_f \leq 1$ volt, $\sigma_{am} \leq 2$ g's, and σ_{um}

116

≤ 3 cm, where σ_f is the RMS actuator input, σ_{am} is the RMS actuator acceleration, σ_{um} is the RMS actuator displacement, and g is the gravitational acceleration.

The next five performance measures (J_6 to J_{10}) are the maximum responses of the structure and actuator subjected to two time-scaled earthquake records, the NS 1940 El Centro record and the NS 1968 Hachinohe record (Figure 8.5). Similar to the first five criteria, they are the maximum relative displacement of floors (J_6), the maximum acceleration of floors (J_7), and the maximum displacement, velocity, and acceleration of the actuator (J_8, J_9, and J_{10}, respectively). Three constraints are included: max $|f| \leq 3$ volts, max $|u_m| \leq 9$ cm, and max $|a_m| \leq 6$ g's, where f is the actuator input, u_m is the actuator displacement, and a_m is the actuator acceleration.

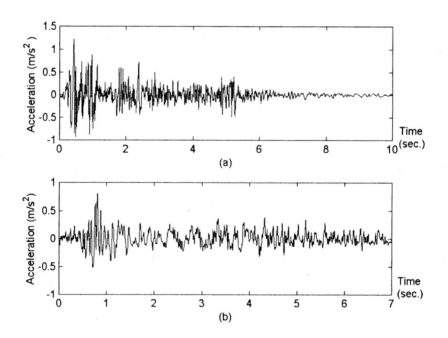

Figure 8.5 Time-scaled time histories used for simulation: (a) NS 1940 El Centro record, (b) NS 1968 Hachinohe record

The performance of the proposed control algorithm for the benchmark problem is demonstrated by numerical simulations using MATLAB SIMULINK (1999). Figure 8.6 shows the SIMULINK diagram for the wavelet-hybrid feedback-LMS control model for the benchmark problem subjected to the time-scaled El Centro earthquake. In this diagram, the filtered-x LMS adaptive controller consists of the "Filtered-x Signal Producer" block and the "LMS Adaptive Controller" block. As illustrated in Figure 8.3, the control force, $f_x(n)$, produced by the filtered-x LMS controller, and feedback control force, $f_b(n)$, produced by the feedback controller are summed to yield the total control force, $f_c(n)$. A two-level low-pass filter using a Daubechies wavelet with 3 vanishing moments is used for this simulation. This two-level wavelet filtering produces a cutoff frequency of about 40 Hz. Thus, the filtered signal covers all three significant natural frequencies of the structural system. Following the sample LQG controller design provided in Spencer et al. (1998), a few selected acceleration responses are used as feedback states for the LQG controller as well as error signals to the filtered-x LMS adaptive controller. The order of the FIR filter is set to 100. The simulation results indicate that choosing a larger order FIR filter would have an insignificant effect on the performance of the proposed control model, while it requires more calculation time.

The number of vanishing moments does not affect the results significantly, even though it affects the computational time required. Simulations using wavelets with a larger number of vanishing moments produce a slight improvement of results due to their better frequency locality but more CPU requirement, while using wavelets with a fewer number of vanishing moments, for example, a Harr wavelet, produces less vibration suppression results due to their poor frequency locality. However, the difference in performance due to a different number of vanishing moments is

118

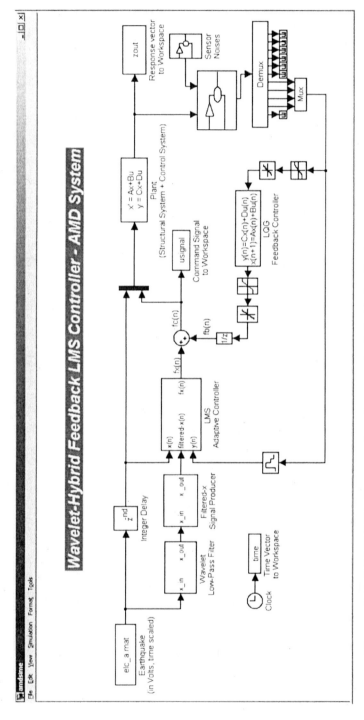

Figure 8.6 SIMULINK diagram for the wavelet-hybrid feedback-LMS control model for the benchmark problem subjected to the time-scaled El Centro earthquake

not significant considering the overall performance improvement over the sample LQG controller.

Figure 8.7 shows uncontrolled and controlled third floor displacement and acceleration of the benchmark structure subjected to the time-scaled El Centro earthquake. These responses are compared to responses controlled by the sample LQG controller presented in Spencer et al., (1998) in Figure 8.8.

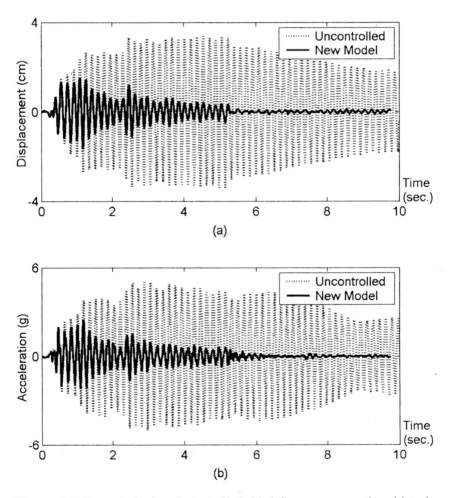

Figure 8.7 Uncontrolled and controlled third floor response time histories of the benchmark structure subjected to the time-scaled El Centro earthquake: (a) displacement, (b) acceleration

120

The values of max $|f|$, max $|u_m|$, and max $|a_m|$ for the El Centro earthquake simulation are 1.061, 3.845, and 5.849, respectively, and satisfy the given constraint.

Figure 8.9 shows uncontrolled and controlled third floor displacement and acceleration time history of the benchmark structure subjected to the time-scaled Hachinohe earthquake. The comparison with the sample LQG

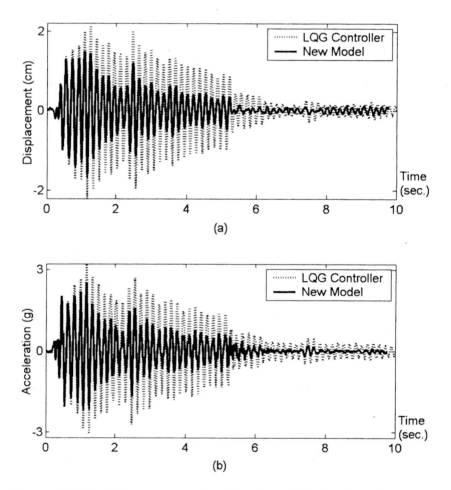

(a)

(b)

Figure 8.8 Third floor response time histories of the benchmark structure subjected to the time-scaled El Centro earthquake using the new control model and the sample LQG controller: (a) displacement, (b) acceleration

controller is presented in Figure 8.10. The values of max $|f|$, max $|u_m|$, and max $|a_m|$ for the Hachinohe earthquake simulation are 1.247, 4.512, and 5.783, respectively.

The simulation using the artificial ground acceleration record (Figure 8.11) with a rather large duration of 300 seconds represents steady state vibrations as opposed to the transient vibration of the time-scaled El Centro

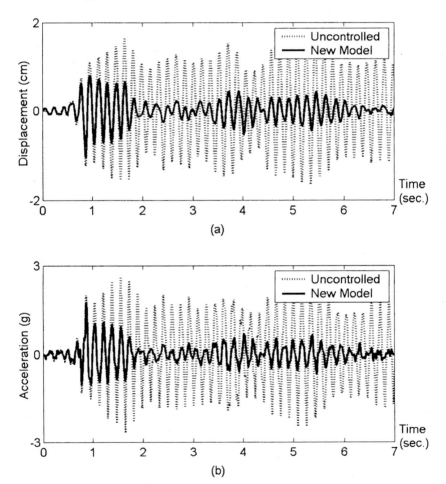

Figure 8.9 Uncontrolled and controlled third floor response time histories of the benchmark structure subjected to the time-scaled Hachinohe earthquake: (a) displacement, (b) acceleration

earthquake record with a duration of 10 seconds (Figures 8.7 and 8.8) and the Hachinohe earthquake record with a duration of 7 seconds (Figures 8.9 and 8.10). From Figures 8.11, it can be observed that structural responses controlled by the wavelet-hybrid feedback-LMS control algorithm resemble those of the LQG control algorithm in the early stage. But, as time goes on

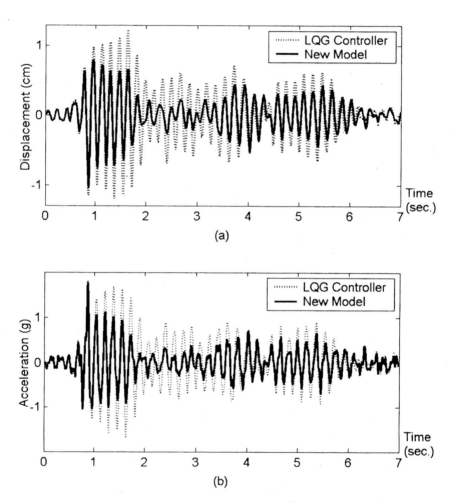

Figure 8.10 Third floor response time histories of the benchmark structure subjected to the time-scaled Hachinohe earthquake using the new control model and the sample LQG controller: (a) displacement, (b) acceleration

the new control model suppresses the vibrations more effectively. Consequently, it can be concluded that both transient and steady state vibrations are effectively controlled by the new control algorithm. For the stochastic signal, σ_f, σ_{am}, and σ_{um} are 0.388, 1.9910, and 1.442, respectively.

Simulation results for the evaluation criteria J_1 to J_{10} for the stochastic signal and the time-scaled El Centro and Hachinohe earthquake records are

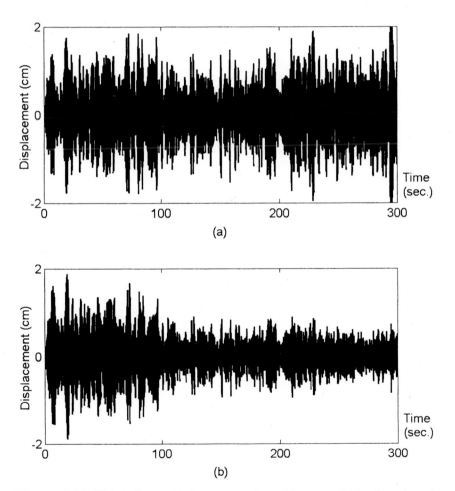

Figure 8.11 Third floor displacement time history of the benchmark structure subjected to the stochastic signal using: (a) sample LQG controller, (b) wavelet-hybrid feedback-LMS control

summarized in Table 8.1. The RMS values of displacement (J_1) and acceleration (J_2) for the stochastic signal using the new control model are 19% and 21%, respectively, less than the corresponding values using the sample LQG model. The maximum values of displacement (J_6) and acceleration (J_7) for the time-scaled El Centro earthquake using the new control model are 20% and 21%, respectively, less than the corresponding values using the sample LQG model. The corresponding reductions for the time-scaled Hachinohe earthquake are 18% for the maximum displacement and 1% for the maximum acceleration. For the simulation using the time-scaled Hachinohe earthquake, another wavelet, the orthogonal Symmlet wavelet with 3 vanishing moments, is applied to show the applicability of other types of orthogonal wavelets. The simulation results presented in Figures 8.9 and 8.10 and Table 8.1 show that the efficacy of the algorithm is not affected noticeably by the choice of wavelets as long as they are orthogonal wavelets.

Table 8.1 Comparison of evaluation criteria for the AMD benchmark problem

Quantities	LQG		Wavelet-hybrid feedback-LMS control	
(a) Stochastic Signal				
J_1	0.283		0.228	
J_2	0.440		0.346	
J_3	0.510		1.101	
J_4	0.513		1.013	
J_5	0.628		1.112	
(b) Time-scaled Earthquake				
	El Centro	Hachinohe	El Centro	Hachinohe
J_6	0.402	0.456	0.320	0.373
J_7	0.636	0.681	0.500	0.679
J_8	0.593	0.669	0.142	1.689
J_9	0.606	0.771	1.146	1.887
J_{10}	0.940	1.280	1.158	2.073

The CPU time on an 800 MHz personal computer using the wavelet-hybrid feedback-LMS control algorithm for the El Centro earthquake simulation was 8.4 seconds compared with 8.2 seconds for the LQG control algorithm. Thus, the additional computational burden due to the introduction of wavelet filtering and LMS filter coefficient adaptation is negligible.

8. 4. Concluding remarks

As demonstrated in Chapter 7, low-pass filtering of dynamic environmental disturbance signals due to winds, earthquakes, and waves is required when the hybrid feedback-LMS algorithm is used for control of real civil structures, because the frequency bandwidths of such environmental signals are much wider than those of common structural systems. In this chapter, it is shown that the wavelet transform can be effectively used as a low-pass filter for the control of civil structures.

The wavelet-hybrid feedback-LMS algorithm proposed in this chapter integrates the wavelet low-pass filter with the hybrid feedback-LMS adaptive controller introduced in Chapter 7. Simulation results demonstrate that the proposed algorithm is effective for control of both steady and transient vibrations without any significant additional computational burden. Both widely used LQR and LQG control algorithms are used for the feedback controller in the examples presented. As such, it is concluded that the proposed control model can be used readily to enhance the performance of existing feedback control algorithms.

Chapter 9

Hybrid Control of 3D Irregular Buildings under Seismic Excitation

9.1. Introduction

A good number of research articles have been published on active, semi-active, and hybrid control of structures subjected to dynamic excitations in recent years. However, most of these articles deal with two dimensional structures or small three dimensional (3D) structures with symmetrical plans. The two dimensional (2D) analysis of torsionally coupled structures often results in underestimation of coupled lateral and torsional responses.

In this chapter, the hybrid damper-TLCD control model, presented in section 2.5, is employed for control of responses of 3D irregular buildings under various seismic excitations. First, the equations of motion for the combined building and TLCD system are derived for multistory building structures with rigid floors and plan and elevation irregularities. Then, optimal control of 3D irregular buildings equipped with a hybrid damper-TLCD system is described and major steps involved are delineated. Next, the wavelet-hybrid feedback-LMS control algorithm, presented in Chapter 8, is applied to find the optimum control forces. Two multistory moment-resisting building structures with vertical and plan irregularities are used to investigate the effectiveness of the new control system in controlling the seismic response of irregular buildings.

9. 2. Analytical model

9.2.1. Coupled dynamic response of 3D irregular buildings

The N-story three-dimensional building model considered in this chapter can have both plan and elevation (setback) irregularities. Floor diaphragms are assumed to be rigid. Horizontal loads are transferred to the columns through the rigid floor diaphragms. In general, the center of mass, C_M, does not coincide with the center of resistance, C_R, in each floor [Figure 9.1(a)]. The centers of mass and resistance of floors do not have to lie on the same vertical axes (their locations can vary from floor to floor). For such buildings, lateral and torsional motions under seismic excitations are coupled. The structural model has three displacement degrees of freedom at each floor level i: translations in the x- and y-directions, u_i and v_i, respectively, and a rotation about the vertical axis z passing through the center of mass, θ_i ($i = 1$, 2, ..., N). The dynamic equation of motions of the 3D building structure under seismic excitations is written as

$$M\ddot{u} + C\dot{u} + Ku = -Mr_g\ddot{u}_g \qquad (9.1)$$

where M, C, K, respectively, are the $3N$ x $3N$ mass, damping, and stiffness matrices of the structure and $\ddot{u}_g$ is the ground acceleration. The displacement vector u and the ground influence vector r_g have the forms

$$u = [u_1^T\ u_2^T\ u_3^T\ ...\ u_N^T]^T \qquad (9.2)$$

$$r_g = [r_{g,1}^T\ r_{g,2}^T\ r_{g,3}^T\ ...\ r_{g,N}^T]^T \qquad (9.3)$$

where

$$\boldsymbol{u}_i = \begin{bmatrix} u_i \\ v_i \\ \theta_i \end{bmatrix}, \qquad i = 1, 2, \ldots, N \tag{9.4}$$

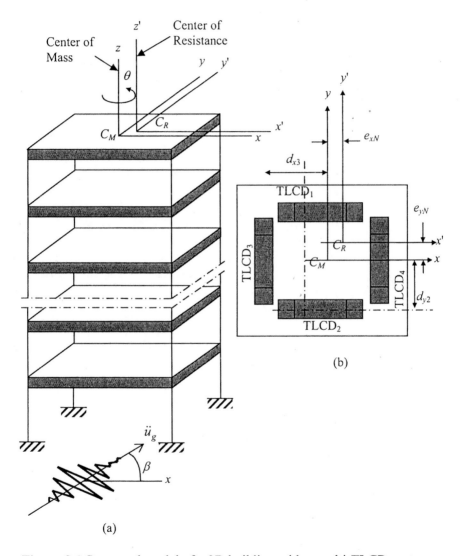

Figure 9.1 Structural model of a 3D building with a multi-TLCD system on the roof subjected to coupled translational and torsional motions

$$
\mathbf{r}_{g,i} = \begin{bmatrix} \cos \beta \\ \sin \beta \\ 0 \end{bmatrix}, \qquad i = 1, 2, \ldots, N \tag{9.5}
$$

and β is the direction angle of the incident earthquake motion measured from the x-axis [Figure 9.1(a)].

9.2.2. Dynamic equation of a TLCD system

In this example, two pairs of TLCDs are installed on the roof of the building, one pair along each principal axis of the building plan [Figure 9.1(b)]. This configuration is selected in order to maximize the vibration suppression and to avoid additional undesirable torsional effects. Referring to Figure 2.2, the equations of motion of each TLCD installed on the roof of the N-story building, in the directions of the x- and y-axes are (Liang et al., 2000)

$$
m_{x_i} \ddot{x}_i(t) + \frac{\rho A_{x_i} \xi_{x_i}(t) |\dot{x}_i(t)|}{2} \dot{x}_i(t) + k_{x_i} x_i(t) = -\alpha_{x_i} m_{x_i} \ddot{u}_N(t) + \alpha_{x_i} m_{x_i} d_{y_i} \ddot{\theta}_N(t) \tag{9.6}
$$

$$
m_{y_i} \ddot{y}_i(t) + \frac{\rho A_{y_i} \xi_{y_i}(t) |\dot{y}_i(t)|}{2} \dot{y}_i(t) + k_{y_i} y_i(t) = -\alpha_{y_i} m_{y_i} \ddot{v}_N(t) - \alpha_{y_i} m_{y_i} d_{x_i} \ddot{\theta}_N(t) \tag{9.7}
$$

where the last terms in the right hand side of Eqs. (9.6) and (9.7) represent the torsional contribution. In Eqs. (9.6) and (9.7) m_{x_i}, m_{y_i}, k_{x_i}, k_{y_i}, α_{x_i}, and α_{y_i} are, respectively, mass, equivalent stiffness, and width-to-length ratio of the liquid tube of the ith TLCD in the x- and y-directions defined by

$$
m_{x_i} = \rho A_{x_i} L_{x_i} \qquad\qquad m_{y_i} = \rho A_{y_i} L_{y_i} \tag{9.8}
$$

$$k_{x_i} = 2\rho g A_{x_i} \qquad\qquad k_{y_i} = 2\rho g A_{y_i} \qquad\qquad (9.9)$$

$$\alpha_{x_i} = {B_{x_i}} \Big/ {L_{x_i}} \qquad\qquad \alpha_{y_i} = {B_{y_i}} \Big/ {L_{y_i}} \qquad\qquad (9.10)$$

and A_{x_i}, A_{y_i}, B_{x_i}, B_{y_i}, L_{x_i}, and L_{y_i} are the cross-sectional area, the width, and the length of the liquid tube of the ith TLCD in the x- and y-directions, respectively; ξ_{x_i} and ξ_{y_i} are the coefficient of head loss determined by the opening ratio of the orifice of the ith TLCD in the x- and y-directions, respectively; x_i is the displacement of the liquid column of the ith TLCD which is parallel to the x-axis; y_i is displacement of the liquid column of the ith TLCD which is parallel to the y-axis; ρ is the density of the liquid; g is the gravitational acceleration; and d_{x_i} and d_{y_i} are the x- and y-coordinates of the center of the ith TLCD in the xy coordinate system with the origin at the center of mass, C_M.

9.2.3. Equations of motion for the combined building and TLCD system

Equations of motion for the combined building and TLCD system are obtained by combining Eqs. (9.1), (9.6) and (9.7). The results in matrix notation are

$$\begin{bmatrix} M + M' & M_{DT} \\ M_{TD} & M_T \end{bmatrix} \begin{Bmatrix} \ddot{u} \\ \ddot{u}_T \end{Bmatrix} + \begin{bmatrix} C & [0]_{3N\times4} \\ [0]_{4\times3N} & C_T \end{bmatrix} \begin{Bmatrix} \dot{u} \\ \dot{u}_T \end{Bmatrix}$$

$$+ \begin{bmatrix} K & [0]_{3N\times4} \\ [0]_{4\times3N} & K_T \end{bmatrix} \begin{Bmatrix} u \\ u_T \end{Bmatrix} = -\begin{Bmatrix} Mr_g \\ M_T r_T \end{Bmatrix} \ddot{u}_g \qquad (9.11)$$

where the matrices

$$M_T = \text{diag}\begin{pmatrix} m_{x_1} & m_{x_2} & m_{y_3} & m_{y_4} \end{pmatrix} \tag{9.12}$$

$$C_T = \text{diag}\left(\frac{\rho A_{x_1} \xi_{x_1}(t)|\dot{x}_1(t)|}{2} \quad \frac{\rho A_{x_2} \xi_{x_2}(t)|\dot{x}_2(t)|}{2} \quad \frac{\rho A_{y_3} \xi_{y_3}(t)|\dot{y}_3(t)|}{2} \quad \frac{\rho A_{y_4} \xi_{y_4}(t)|\dot{y}_4(t)|}{2} \right)$$
$$\tag{9.13}$$

$$K_T = \text{diag}\begin{pmatrix} k_{x_1} & k_{x_2} & k_{y_3} & k_{y_4} \end{pmatrix} \tag{9.14}$$

are the mass, equivalent damping, and equivalent stiffness matrices of the TLCD system; $u_T = \begin{bmatrix} x_1 & x_2 & y_3 & y_4 \end{bmatrix}^T$ is the vector containing the vertical displacements of the liquid in the four TLCDs. The ground influence vector associated with TLCDs is represented by r_T as

$$r_T = \begin{bmatrix} \cos\beta & \cos\beta & \sin\beta & \sin\beta \end{bmatrix}^T \tag{9.15}$$

The mass coupling matrices M_{DT} and M_{TD} and the mass contribution of TLCDs to the structural mass matrix represented by M' are

$$M_{DT} = \begin{bmatrix} [0]_{(3N-3)\times 4} \\ \begin{bmatrix} \alpha_{x_1} m_{x_1} & \alpha_{x_2} m_{x_2} & 0 & 0 \\ 0 & 0 & \alpha_{y_3} m_{y_3} & \alpha_{y_4} m_{y_4} \\ -\alpha_{x_1} m_{x_1} d_{y_1} & -\alpha_{x_2} m_{x_2} d_{y_2} & \alpha_{y_3} m_{y_3} d_{x_3} & \alpha_{y_4} m_{y_4} d_{x_4} \end{bmatrix}_{3\times 4} \end{bmatrix} \tag{9.16}$$

$$M_{TD} = M_{DT}^T \tag{9.17}$$

$$M' = \begin{bmatrix} [0]_{(3N-3)\times(3N-3)} & [0]_{(3N-3)\times 3} \\ [0]_{3\times(3N-3)} & \begin{bmatrix} M'_{11} & 0 & 0 \\ 0 & M'_{22} & 0 \\ 0 & 0 & M'_{33} \end{bmatrix}_{3\times 3} \end{bmatrix} \tag{9.18}$$

where

$$M'_{11} = m_{x_1} + m_{x_2} + m_{y_3} + m_{y_4} \tag{9.19}$$

$$M'_{22} = M'_{11} \tag{9.20}$$

$$M'_{33} = I_{x_1} + I_{x_2} + I_{y_3} + I_{y_4} \tag{9.21}$$

in which I_{x_1} ($= m_{x_1} d^2_{y_1}$) and I_{x_2} are the inertia moments of the liquid in TCLD$_1$ and TLCD$_2$ in the x-direction relative to the mass center of the roof floor (I_{x_1}); and I_{y_3} and I_{y_4} are the inertia moments of the liquid in TCLD$_3$ and TLCD$_4$ in the y-direction relative to the mass center of the roof floor [Figure 9.1(b)].

9. 3. Optimal control of 3D irregular buildings equipped with a hybrid damper-TLCD system

The wavelet-hybrid feedback-LMS control algorithm, presented in Chapter 8, is employed to find optimum control forces. The following presents the major steps involved in the wavelet-based optimal control of 3D irregular buildings equipped with a hybrid damper-TLCD system.

Step 1. Construct a finite element model of the building and obtain the dynamic characteristics of the building structure such as natural frequencies and mode shapes.

Step 2. Determine the design parameters of semi-active TLCD (mass ratio, tuning ratio, and the maximum value of the head loss coefficient) and passive dampers (required damping ratios and configuration). The detailed procedure involved in the determination of the design parameters is summarized in section 2.5.1.

Step 3. Design the wavelet-based low-pass filter. This includes the selection of the family of the wavelet, determination of the number of the vanishing moment for the selected wavelet, and the level of filtering. For fast and real time implementation the wavelet needs to be orthogonal. Wavelets with larger numbers of vanishing moments produce better frequency locality and therefore better controllability but require more computer processing power. On the other hand, wavelets with fewer numbers of vanishing moments produce less vibration suppression due to their poor frequency locality. The level of wavelet low-pass filtering is chosen such that the cutoff frequency is greater than the largest significant natural frequency of the structure.

Step 4. Design the feedback controller. For the feedback controller in the control algorithm, either the LQR or the LQG algorithm (Soong, 1990; Adeli and Saleh, 1997, 1999; Christenson et al., 2003; Connor, 2003) can be used.

Step 5. Integrate the feedback controller (LQR or LQG) with the filtered-x LMS controller and estimate the control force-to-output relationship of the system using the FIR filter in the offline LMS implementation.

Step 6. Construct the optimal controller by integrating the feedback-LMS controller created in Step 5 with the wavelet low-pass filter designed in Step 3. The wavelet low-pass filter must be arranged such that the wavelet filtering of the external excitation affects only the filtered-x LMS adaptive controller and not the feedback controller. This is because the wavelet low-pass filter is primarily used for eliminating higher frequency components of the external excitation which impede the stabilization of the FIR filter coefficients. Further, the input to the feedback controller needs to be the response of the structural system subjected to unfiltered signals.

9. 4. Examples

Two multistory moment-resisting building structures are investigated in this chapter, representing two types of irregular building configurations – plan and vertical irregularities – as defined in the International Building Code (IBC, 2000). They are designed according to the American Institute of Steel Construction (AISC) Load and Resistance Factor Design (LRFD) specifications (AISC, 1998) for the combination of static dead and live loads and the lateral loads obtained by the equivalent linear static load procedure described in IBC (2000).

For dynamic analysis, the building structures are modeled with finite elements. Columns and beams are modeled as three-dimensional frame elements with two end nodes. The floor slab is modeled with four-node plane elements. The floor elements are used for generating the floor mass only and their stiffness contributions are ignored due to the rigid diaphragm modeling assumption mentioned earlier. Each node has six (three displacements and three rotations) DOFs. The same three simulated earthquake ground accelerations used in section 2.2.4 are employed.

9.4.1. 12-story moment-resisting space steel frame with vertical irregularity

This is a 12-story moment-resisting steel frame with a vertical setback on the fifth floor and a height of 54 m as shown in Figure 9.2. The example was first introduced in the literature by Adeli and Saleh (Saleh and Adeli, 1998b; Adeli and Saleh, 1999) for study of active control of structures. The same geometry and static loadings are employed here. The structure has 148 members, 77 nodes, and 462 DOFs prior to applying boundary conditions, rigid diaphragm constraints, and the dynamic condensation. Applying boundary conditions and rigid diaphragm constraints results in 240 DOFs.

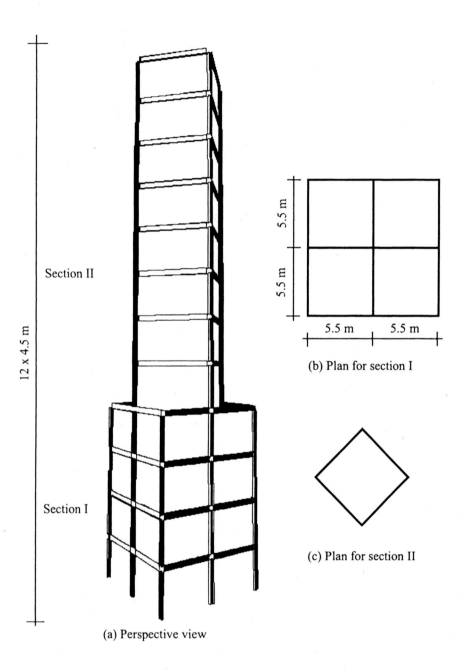

(a) Perspective view

(b) Plan for section I

(c) Plan for section II

Figure 9.2 Example 1: Twelve-story moment-resisting steel frame with a vertical setback (adapted from Adeli and Saleh, 1999)

They are further reduced to 36 DOFs by the Guyan reduction of vertical DOFs and rotational DOFs about two horizontal axes (Craig, 1981).

The static loading on the building consists of uniformly distributed floor dead and live loads of 2.88 kPa (60 psf) and 2.38 kPa (50 psf), respectively. A total lateral force (base shear) of 243 kN is obtained and distributed over the height of the structure using the equivalent linear static load approach provided by IBC (2000). Each floor shear force is distributed to the nodes in that floor in proportion to nodal masses.

A damping ratio of 2% is used for each mode. The first five mode shapes of this example are presented in Figure 9.3. The shape of the first mode with a frequency of 0.564 Hz is almost identical to the second mode shape with a frequency of 0.583 Hz except for their directions because the building is symmetric in plan. Thus, the first two mode shapes are shown in one figure [Figure 9.3(a)]. Similarly, the shapes of the fourth and fifth modes are almost identical except for their directions and therefore are presented in one figure [Figure 9.3(c)]. Even though the building is symmetric in plan, the story stiffnesses are different in two principal directions because columns are wide-flange shapes with unequal cross-sectional moments of inertia with respect to their principal axes. Thus, the centers of mass and resistance (or rigidity) in each floor do not coincide, resulting in coupling of torsional and lateral vibrations of the building.

Figure 9.4 shows top floor displacements of the structure subjected to EQ-I as a function of the angle of incidence of the ground acceleration (β) in the range of -90° to 90°. Figures 9.4(a) and 9.4(b) show the maximum displacement in the x- and y-directions, respectively. Figure 9.4(c) shows the largest RMS value of displacements in the x- and y- directions. The coupling effect of lateral and torsional vibrations is observed in Figure 9.4. If there were no coupling, the maximum displacements in the x- and y-directions

(a)   (b)   (c)

Figure 9.3 Mode shapes for the twelve-story structure: (a) modes 1 (frequency = 0.564 Hz) and 2 (frequency = 0.583 Hz), (b) mode 3 (frequency = 0.690 Hz), (c) modes 4 (frequency = 1.25 Hz) and 5 (frequency = 1.30 Hz)

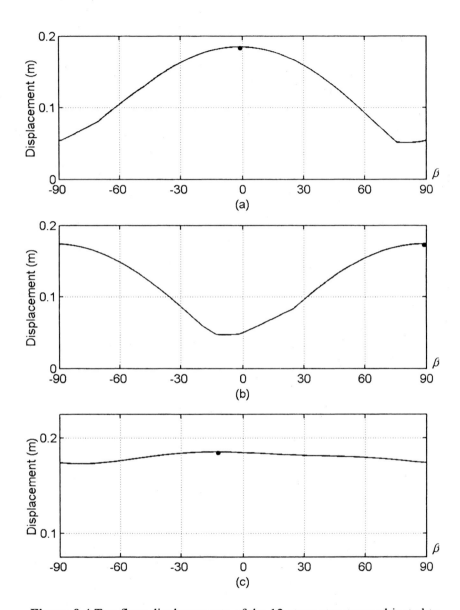

Figure 9.4 Top floor displacements of the 12-story structure subjected to EQ-I as a function of the angle of incidence of the ground acceleration: (a) maximum displacement in the x-direction, (b) maximum displacement in the y-direction, (c) the largest RMS value of displacements in the x- and y-directions

would be at $\beta = 0°$ and $90°$, respectively. However, as noted in Figure 9.4, the maximum displacements in the x- and y-directions occur at $\beta = -1.7°$ and $88.1°$, respectively. These values are near $0°$ and $90°$, respectively, because the structure is symmetric in plan and the centers of mass and resistance are close to each other. The incidence angles that produce the largest displacements are identified with bullets in Figure 9.4. The largest RMS value of the x- and y-displacements of the top floor occurs at $\beta = -13.0°$ even though the maximum displacements in the x- and y-directions occur near $0°$ and $90°$, respectively. The 2D dynamic analysis of this building in the x- or y-direction underestimates the maximum response of the structure by up to 4%. Thus, a 3D dynamic analysis needs to be performed in order to obtain more accurate results.

To design a hybrid damper-TLCD control system for a 3D building structure, two parallel sets of TLCDs are used, as noted earlier and shown in Figure 9.1. The same TLCD unit is used in each direction. But, different TLCDs with different parameters are used in perpendicular directions. For the x-direction the following values are used for the TLCD parameters: mass ratio $\mu = 0.02$, tuning ratio $f = 0.975$, maximum head loss coefficient ξ_{max}, = 30, and liquid tube width-to-length ratio $\alpha = 0.9$. In this book, viscous fluid dampers are used in the form of diagonal or Chevron bracings but without providing any additional stiffness (Hanson and Soong, 2001). Dampers are chosen such that the damping ratio for the fundamental mode of the structure including its intrinsic damping is increased to 5 percent.

Figure 9.5 shows the time histories of the top floor displacement of the structure subjected to EQ-I in the x-direction using three controlled systems: (a) passive damper system, (b) semi-active TLCD system, (c) hybrid damper-TLCD system. These responses are obtained when the earthquake motion is applied in the x-direction ($\beta = 0°$) and assuming only the two TLCDs parallel

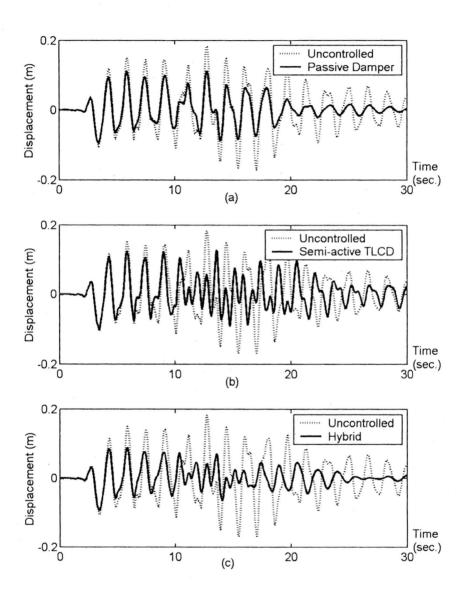

Figure 9.5 Time histories of the top floor displacement of the 12-story structure subjected to EQ-I in the *x*-direction using three controlled systems: (a) passive damper system, (b) semi-active TLCD system, (c) hybrid damper-TLCD system

to the x-axis are installed. It is observed from Figure 9.5 that the combination of the passive damper and semi-active TLCD systems reduces the response substantially and maximizes the control performance by acting complementary to each other. Further, the integration of a passive supplementary damping system with a semi-active TLCD system provides increased reliability and maximum operability during power failure as

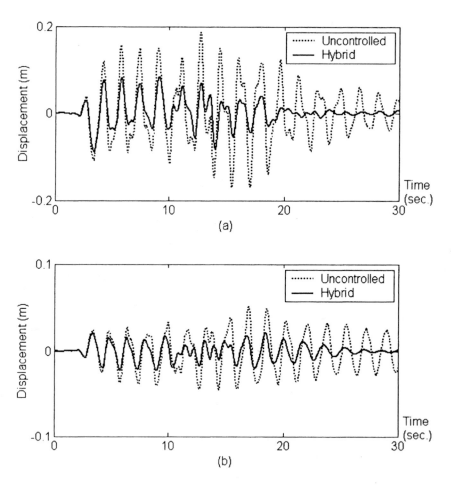

Figure 9.6 Top floor displacement responses of the 12-story structure subjected to EQ-I with $\beta = -13.0°$ using the hybrid damper-TLCD system: (a) in the x-direction, (b) in the y-direction

described in Chapter 2.

For the 3D control of the structure, values of the damping coefficient for supplementary dampers and the design parameters for the TLCD system in the y-direction are chosen similar to those chosen for the x-direction because the structure is doubly symmetric in plan. The resulting top floor displacement responses of the example structure subjected to EQ-I with $\beta =$ -13.0° in the x- and y-directions using the hybrid damper-TLCD system are presented in Figure 9.6. The earthquake incidence angle of $\beta = $ -13.0° is used because it produces the largest RMS value of displacements in the x- and y-directions in the simulation presented in Figure 9.4. Compared with the uncontrolled system, the hybrid damper-TLCD control system reduces the maximum displacement in the x- and y-directions by 53% and 56%, respectively.

The largest RMS acceleration and displacement responses of the top floor subjected to all three simulated earthquake ground accelerations with β = -13.0° are presented in Table 9.1. Compared with the uncontrolled system, the hybrid damper-TLCD control system reduces the RMS displacement by 46-50% and RMS acceleration by 61-71%.

Table 9.1 The largest RMS responses of the top floor of the 12-story structure subjected to simulated earthquake ground accelerations EQ-I, EQ-II, and EQ-III with $\beta = $ -13.0°

Earthquake	Response	Uncontrolled	Hybrid damper-TLCD controlled
EQ-I	Displacement	4.18 cm	2.10 cm
	Acceleration	1.52 m/sec^2	0.59 m/sec^2
EQ-II	Displacement	5.86 cm	3.17 cm
	Acceleration	2.56 m/sec^2	0.75 m/sec^2
EQ-III	Displacement	7.43 cm	3.93 cm
	Acceleration	3.83 m/sec^2	1.33 m/sec^2

9.4.2. 8-story moment-resisting space steel frame with plan irregularity

This is an 8-story moment-resisting steel frame with a plan irregularity and a height of 36 m created in this study and shown in Figure 9.7. The structure has 208 members, 99 nodes, and 594 DOFs prior to applying boundary conditions, rigid diaphragm constraints, and the dynamic condensation. Applying boundary conditions and rigid diaphragm constraints results in 288 DOFs. They are further reduced to 24 DOFs by the Guyan reduction of vertical DOFs and rotational DOFs about two horizontal axes.

The static loading on the building consists of uniformly distributed floor dead and live load of 4.78 kPa (100 psf) and 3.35 kPa (70 psf), respectively. A total lateral force (base shear) of 963 kN is obtained and distributed over the height of the structure using the equivalent linear static load approach provided by IBC (2000). Each floor shear force is distributed to the nodes in that floor in proportion to nodal masses.

Because of plan irregularity substantially more translational and torsional coupling effect is expected in this example compared with the previous example. Figure 9.8 shows the top floor displacements of the first three modes of vibrations: (a) mode 1 with a frequency of 0.57 Hz, (b) mode 2 with a frequency of 0.72 Hz, (c) mode 3 with a frequency of 0.75 Hz (Displacements in this figure are magnified by 5). Figure 9.9 shows top floor displacements of the structure subjected to EQ-I as a function of the angle of incidence of the ground acceleration (β) in the range of -90° to 90°. Figures 9.9(a) and 9.9(b) show the maximum displacement in the x- and y-directions, respectively. Figure 9.9(c) shows the largest RMS value of displacements in the x- and y- directions.

Substantial coupling effect of lateral and torsional vibrations is observed in Figure 9.9. The incidence angles that produce the largest displacements are identified with bullets. The maximum displacements in the

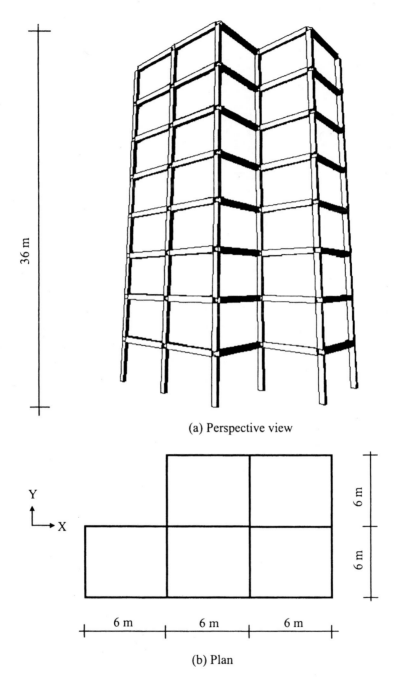

(a) Perspective view

(b) Plan

Figure 9.7 Example 2: Eight-story moment-resisting steel frame with plan irregularity

146

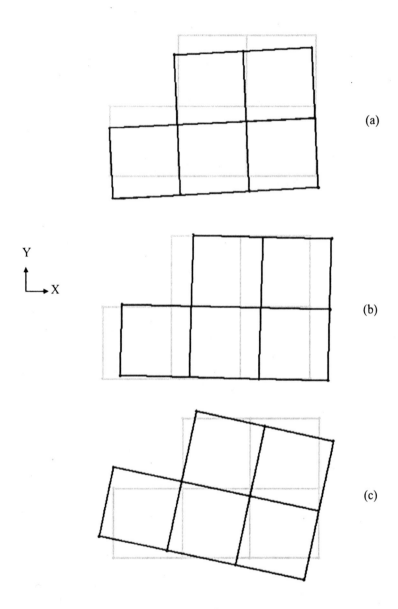

Figure 9.8 Top floor displacements of the first three modes of vibrations for the 8-story structure: (a) mode 1 (frequency = 0.57 Hz), (b) mode 2 (frequency = 0.72 Hz), (c) mode 3 (frequency = 0.75 Hz)

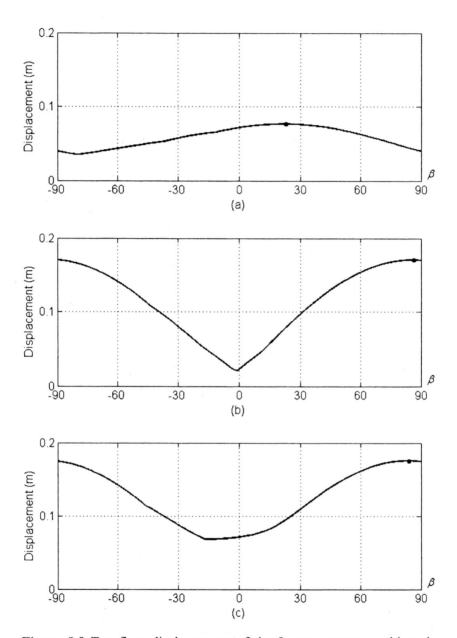

Figure 9.9 Top floor displacements of the 8-story structure subjected to EQ-I as a function of the angle of incidence of the ground acceleration: (a) maximum displacement in the x-direction, (b) maximum displacement in the y-direction, (c) the largest RMS value of displacements in the x- and y-directions

148

and y-directions occur at 22.4° and 85.6°, respectively. The largest RMS value of the x- and y-displacements of the top floor occurs at $\beta = 83.4°$. The 2D dynamic analysis of this building in the x- or y-direction underestimates the maximum response of the structure by up to 7%. Thus, a 3D dynamic analysis needs to be performed in order to obtain accurate results.

The top floor displacement responses of the structure subjected to EQ-I

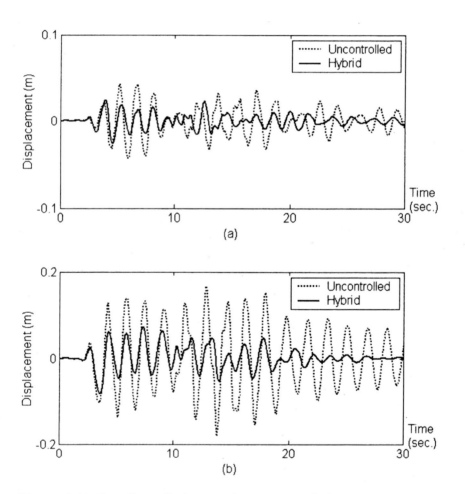

Figure 9.10 Top floor displacement responses of the 8-story structure subjected to EQ-I with $\beta = 83.4°$ using the hybrid damper-TLCD system: (a) in the x-direction, (b) in the y-direction

with $\beta = 83.4°$ in the x- and y-directions using the hybrid damper-TLCD system are presented in Figure 9.10. The earthquake incidence angle of $\beta = 83.4°$ is used because it produces the largest RMS value of displacements in the x- and y-directions in the simulation presented in Figure 9.9. Compared with the uncontrolled system, the hybrid damper-TLCD control system reduces the maximum displacement in the x- and y-directions by 38% and 54%, respectively.

The largest RMS acceleration and displacement responses of the top floor subjected to all three simulated earthquake ground accelerations with $\beta = 83.4°$ are presented in Table 9.2. Compared with the uncontrolled system, the hybrid damper-TLCD control system reduces the RMS displacement by 56-67% and RMS acceleration by 71-84%.

Table 9.2 The largest RMS responses of the top floor of the 8-story structure subjected to simulated earthquake ground accelerations EQ-I, EQ-II, and EQ-III with $\beta = 83.4°$

Earthquake	Response	Uncontrolled	Hybrid damper-TLCD controlled
EQ-I	Displacement	4.28 cm	1.85 cm
	Acceleration	1.18 m/sec^2	0.34 m/sec^2
EQ-II	Displacement	6.98 cm	3.11 cm
	Acceleration	2.44 m/sec^2	0.52 m/sec^2
EQ-III	Displacement	11.1 cm	3.70 cm
	Acceleration	7.55 m/sec^2	1.24 m/sec^2

9. 5. Concluding remarks

In this chapter, the hybrid damper-TLCD control system presented in section 2.5 was empoyed for the control of 3D coupled irregular buildings under various seismic excitations using two multistory moment-resisting building structures with vertical and plan irregularities. The coupled equations of motion for the combined building and TLCD system are derived. Major steps involved in the wavelet-based optimal control of 3D irregular buildings equipped with a hybrid damper-TLCD system are delineated. Two pairs of parallel TLCDs placed along two principal directions of the structural plan are used to control the coupled lateral and torsional response of irregular multistory buildings.

Results of Tables 9.1 and 9.2 and Figures 9.6 and 6.10 clearly indicate that the hybrid damper-TLCD control system can significantly reduce the displacement as well as acceleration responses of 3D irregular buildings subjected to various earthquake ground motions. The same levels of response reduction are achieved for structures with plan and vertical irregularities.

Chapter 10

Vibration Control of Highrise

Buildings under Wind Loading

10. 1. Introduction

In this chapter, the effectiveness of both the semi-active TLCD system and the hybrid damper-TLCD control system, presented in sections 2.5 and 9.4, is investigated for the control of wind-induced motion of highrise buildings. Simulations are performed on the 76-story building benchmark control problem created by Yang et al. (2000) and described briefly in the next section. To evaluate the effectiveness of the control system against wind loading, wind loads obtained from wind tunnel tests (Yang et al., 2000) and the stochastic wind loads defined by the Davenport cross-power spectral density matrix (Yang et al., 1998) are used. The performances of semi-active TLCD and hybrid damper-TLCD control systems are compared with that of a sample ATMD system presented in Yang et al. (2000).

10. 2. 76-story benchmark building

The 76-story benchmark building is a 306-m high office tower with a height-to-width ratio of 7.3 proposed for the city of Melbourne, Australia. The plan of the structure is square with two cut corners (Figure 10.1). The building is a reinforced concrete building consisting of a concrete core and an exterior concrete frame. The typical story height is 3.9 m with the exception of the first floor, which has a height of 10 m, and stories 38 to 40 and 74 to 76,

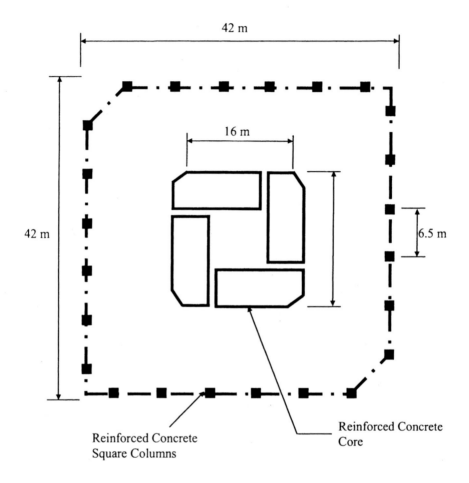

Figure 10.1 The plan sketch of the 76-story benchmark control problem

which have a height of 4.5 m. The building has a total mass including heavy machinery in the plant rooms of 153,000 metric tons. Structural analysis is performed in two dimensions based on the symmetric nature of the plan. The first three natural frequencies of the structure based on a two-dimensional structural analysis are 0.16 Hz, 0.765 Hz, and 1.992 Hz.

To evaluate the effectiveness of a control system against wind loading, wind force data obtained from wind tunnel tests are used (Yang et al., 2000).

The results of the wind tunnel test are for a building model scale of 1:400 and a velocity scale of 1:3. From the data obtained, the first 900 seconds (15 minutes) of wind pressure data are used for the benchmark problem in this study. Figure 10.2 shows the first 5 minutes time histories of resulting wind loads on the 66th and 70th floors as examples.

As a sample control system, an ATMD system with a mass of 500 metric tons installed on the top floor is used. This represents about 45% of the top floor mass and 0.327% of the total mass of the building. The undamped natural frequency and the damping ratio of ATMD are set to 0.16 Hz and 20%, respectively. Per Yang et al. (2000), the ATMD system is designed such that the peak and RMS floor accelerations are less than 15

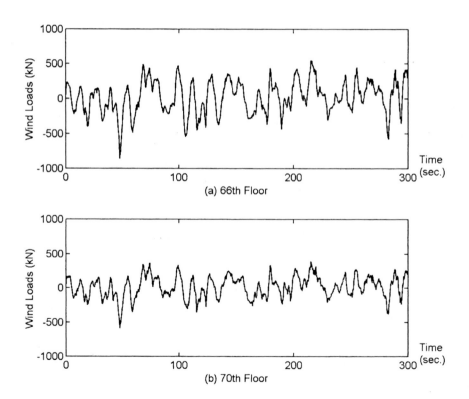

Figure 10.2 Time histories of wind tunnel test loads acting on the 66th and 70th floors

154

cm/s² and 5 cm/s², respectively, considered as maximum allowable values for office buildings.

Twelve criteria, denoted by J_1 to J_{12}, proposed by Yang et al. (2000) evaluate control systems. Smaller numbers for each criterion represent a more effective response control performance. The first six performance measures (J_1 to J_6) are RMS responses of the selected floors of the structure and actuator. The next six performance measures (J_7 to J_{12}) are the peak responses of the selected floors of the structure and actuator. Among the 12 criteria, only 8 criteria (J_1 to J_4 and J_7 to J_{10}) are used in this study, because the other 4 criteria (J_5, J_6, J_{11}, and J_{12}) represent the performance of the actuator. Neither the semi-active TLCD nor the hybrid damper-TLCD system requires any actuator to operate, thus eliminating the need to use the other four criteria.

The first four criteria in terms of RMS responses are

$$J_1 = \max\left(\sigma_{\ddot{x}1},\ \sigma_{\ddot{x}30},\ \sigma_{\ddot{x}50},\ \sigma_{\ddot{x}55},\ \sigma_{\ddot{x}60},\ \sigma_{\ddot{x}65},\ \sigma_{\ddot{x}70},\ \sigma_{\ddot{x}75}\right)/\sigma_{\ddot{x}75o} \quad (10.1)$$

$$J_2 = \frac{1}{6}\sum_i (\sigma_{\ddot{x}i}/\sigma_{\ddot{x}io}) \qquad \text{for } i = 50, 55, 60, 65, 70 \text{ and } 75 \quad (10.2)$$

$$J_3 = \sigma_{x76}/\sigma_{x76o} \qquad \text{for } i = 50, 55, 60, 65, 70, 75 \text{ and } 76 \quad (10.3)$$

$$J_4 = \frac{1}{7}\sum_i (\sigma_{xi}/\sigma_{xio}) \qquad \text{for } i = 50, 55, 60, 65, 70, 75 \text{ and } 76 \quad (10.4)$$

where $\sigma_{\ddot{x}i}$ is the RMS acceleration of the ith floor; $\sigma_{\ddot{x}75o}$ is the RMS acceleration of the 75th floor without control which is equal to 9.142 cm/sec²; $\sigma_{\ddot{x}io}$ is the RMS acceleration of the ith floor without control; σ_{xi} and σ_{xio} are the RMS displacements of the ith floor with and without control, respectively; σ_{x76o} is the RMS displacement of the 76th floor of the uncontrolled building which is equal to 10.137 cm. The values for RMS responses without control are given in the second and third columns of Table 10.1.

The next four criteria in terms of peak responses are

$$J_7 = \max\left(\ddot{x}_{p1}, \ \ddot{x}_{p30}, \ \ddot{x}_{p50}, \ \ddot{x}_{p55}, \ \ddot{x}_{p60}, \ \ddot{x}_{p65}, \ \ddot{x}_{p70}, \ \ddot{x}_{p75}\right)/\ddot{x}_{p75o} \qquad (10.5)$$

$$J_8 = \frac{1}{6}\sum_i \left(\ddot{x}_{pi}/\ddot{x}_{pio}\right) \qquad \text{for } i = 50, 55, 60, 65, 70 \text{ and } 75 \qquad (10.6)$$

$$J_9 = x_{p76}/x_{p76o} \qquad (10.7)$$

$$J_{10} = \frac{1}{7}\sum_i \left(x_{pi}/x_{pio}\right) \qquad \text{for } i = 50, 55, 60, 65, 70, 75 \text{ and } 76 \quad (10.8)$$

where $\ddot{x}_{pi}$ and $\ddot{x}_{pio}$ are the peak acceleration of ith floor with and without control, respectively; x_{pi} and x_{pio} are the peak displacements of ith floor with and without control, respectively; x_{p76o} is the peak displacement of the 76th floor without control which is equal to 32.30 cm and $\ddot{x}_{p75o}$ is the peak acceleration of the 75th floor without control which is equal to 30.33 cm/sec^2. The values for peak responses without control are given in the second and third columns of Table 10.2.

10. 3. Semi-active TLCD system

When a multi-degree of freedom (MDOF) system with a passive TLCD system (Figure 10.3) is subjected to dynamic wind loading, the equations of motion are

$$\begin{bmatrix} M + M' & M_{ST} \\ M_{TS} & m_T \end{bmatrix} \begin{Bmatrix} \ddot{u}(t) \\ \ddot{u}_T(t) \end{Bmatrix} + \begin{bmatrix} C & [0]_{mx1} \\ [0]_{1xm} & c(t) \end{bmatrix} \begin{Bmatrix} \dot{u}(t) \\ \dot{u}_T(t) \end{Bmatrix}$$

$$+ \begin{bmatrix} K & [0]_{mx1} \\ [0]_{1xm} & 2\rho Ag \end{bmatrix} \begin{Bmatrix} u(t) \\ u_T(t) \end{Bmatrix} = - \begin{Bmatrix} F(t) \\ 0 \end{Bmatrix}$$

(10.9)

where mass coupling matrices M_{ST} and M_{TS} and the mass contribution of TLCD to the primary system mass matrix represented by M' are given in section 2.2.4, and $c(t)$ is the damping coefficient of TLCD determined by the head loss coefficient $\xi(t)$ as defined in Eq. (2.14).

For an MDOF structure with a semi-active TLCD system subjected to dynamic wind loading, the the damping coeffiecient $c(t)$ in Eq. (10.9) can be changed by a controllable orifice, and Eq. (10.9) can be rewritten with an additional term on the right-hand side as

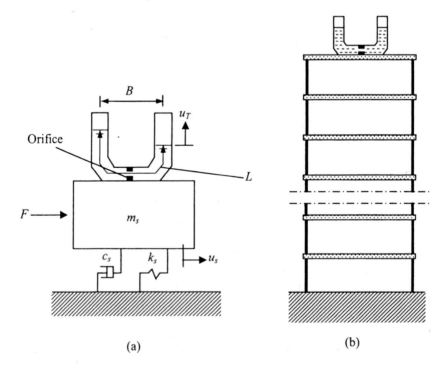

(a)   (b)

Figure 10.3 TLCD system: a) an SDOF system with a TLCD system, b) a TLCD system installed on the roof of a multistory building structure

$$
\begin{bmatrix} M + M' & M_{ST} \\ M_{TS} & m_T \end{bmatrix} \begin{Bmatrix} \ddot{u}(t) \\ \ddot{u}_T(t) \end{Bmatrix} + \begin{bmatrix} C & [0]_{m \times 1} \\ [0]_{1 \times m} & 0 \end{bmatrix} \begin{Bmatrix} \dot{u}(t) \\ \dot{u}_T(t) \end{Bmatrix}
$$
$$
+ \begin{bmatrix} K & [0]_{m \times 1} \\ [0]_{1 \times m} & 2\rho Ag \end{bmatrix} \begin{Bmatrix} u(t) \\ u_T(t) \end{Bmatrix} = - \begin{Bmatrix} F(t) \\ 0 \end{Bmatrix} + \begin{Bmatrix} [0]_{m \times 1} \\ 1 \end{Bmatrix} f_c(t)
$$
(10.10)

where $f_c(t)$ is a semi-active control force of TLCD as defined in Eq. (2.19).

For the numerical simulation of both passive and semi-active TLCD systems, the same mass of 500 tons as the mass of sample ATMD provided by Yang et al. (2000) is used. The optimum tuning ratio is calculated as 0.974 following Yalla and Kareem (2000). The head loss coefficient ξ of 30 is used for the passive TLCD system, and the values of minimum and maximum head loss coefficients, ξ_{min} and ξ_{max}, for the semi-active TLCD system are set to be 0 and 50, respectively. The optimal value of ξ for the passive TLCD system is generally determined based on the statistical RMS value computed for an assumed external excitation. Then, the value of ξ_{max} for the semi-active system used in the proposed hybrid system should be greater than the value of constant ξ for the passive TLCD system. The value of ξ_{min} in this simulation is set to be zero because the head loss coefficient cannot have a negative value practically. The procedure for selection of these numbers is described in section 2.2.4 as part of the general model for design of the proposed hybrid damper-TLCD system.

Optimal values of the head loss coefficients for the semi-active control system are obtained using the wavelet-based optimal control algorithm presented in Chapter 8. In the wavelet-based optimal control algorithm, the wavelet low-pass filter is combined with a classic feedback control algorithm such as the LQR or LQG. The wavelet low-pass filter based on the wavelet multireslution analysis is used for the low-pass filtering of external dynamic excitations, and filtered information is fed into the filtered-x LMS (least mean square) controller to update the filter coefficient. Since the external excitation is included in the calculation of optimal control force, the wavelet-based optimal control algorithm is effective in control of both steady-state and transient vibration.

For the control of wind-induced motion, however, the external excitation consists of more than one component unlike the earthquake-induced motion. The inclusion of every component of external wind loads in the calculation of optimal control force is impractical, and the appropriate selection of wind load for the low-pass filtering is required. In this work, the force acting on the first modal mass is chosen to be included in the control force calculation, since the secondary mass type damper such as TMD and TLCD is generally tuned for the first mode. For the structural system without the secondary mass, the force acting on the first mode, f_1, can be obtained as

$$f_1 = \varphi_1^T f \tag{10.11}$$

where φ_1 is the first mode shape vector.

Figure 10.4 shows the comparison of the displacement frequency responses of the 75th floor for the uncontrolled, passive TLCD, and semi-active TLCD systems in the region of the first three natural frequencies of the primary structure. In this figure, the frequency responses of the uncontrolled and passive TLCD systems near the second and third natural frequencies of the primary system virtually coincide and therefore are indistinguishable. On the other hand, the semi-active TLCD system reduces the responses at every natural frequency. This is because the value of the head loss coefficient is tuned and fixed for the fundamental frequency of the building in the passive TLCD system, while the value varies optimally according to the frequency content of the external force in the semi-active TLCD system.

The RMS displacement and acceleration responses of the selected floors for the sample ATMD and passive and semi-active TLCD systems are presented in Table 10.1 along with the results for the uncontrolled structure. The corresponding peak responses are presented in Table 10.2. It is observed that significant improvement is made when the TLCD system operates semi-actively. While the peak and RMS values of 75th floor accelerations are greater than the maximum allowable values, 15 cm/s^2 and 5 cm/s^2, by 28% and 4%, respectively, for the passive TLCD system, the corresponding values for the semi-active TLCD system are 6% and 18% less than the maximum allowable values, respectively.

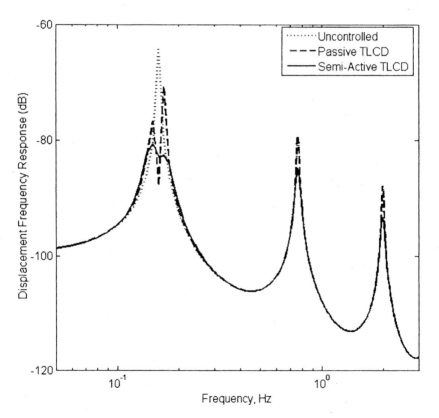

Figure 10.4 Displacement frequency responses of the 75th floor for the uncontrolled, passive TLCD, and semi-active TLCD systems

Table 10.1 Comparison of RMS displacement and acceleration responses of the 76-story building

Floor No.	No control		ATMD		Passive TLCD		Semi-Active TLCD	
	Disp. (cm)	Acc. (cm/s^2)	Disp. (cm)	Acc. (cm/s^2)	Disp. (cm)	Acc. (cm/s^2)	Disp. (cm)	Acc. (cm/s^2)
1	0.02	0.06	0.01	0.06	0.01	0.06	0.01	0.06
30	2.15	2.02	1.26	0.89	1.39	1.15	1.26	0.97
50	5.22	4.78	3.04	2.03	3.36	2.52	3.03	2.08
55	6.11	5.59	3.55	2.41	3.93	2.90	3.54	2.41
60	7.02	6.42	4.08	2.81	4.51	3.28	4.07	2.73
65	7.97	7.31	4.62	3.16	5.11	3.78	4.61	3.13
70	8.92	8.18	5.17	3.38	5.72	4.31	5.16	3.52
75	9.92	9.14	5.74	3.34	6.36	5.17	5.72	4.11
76	10.14	9.35	5.86	4.70	6.50	4.62	5.85	4.01

Table 10.2 Comparison of peak displacement and acceleration responses of the 76-story building

Floor No.	No control		ATMD		Passive TLCD		Semi-Active TLCD	
	Disp. (cm)	Acc. (cm/s^2)	Disp. (cm)	Acc. (cm/s^2)	Disp. (cm)	Acc. (cm/s^2)	Disp. (cm)	Acc. (cm/s^2)
1	0.05	0.22	0.04	0.23	0.04	0.22	0.04	0.23
30	6.84	7.14	5.14	3.38	5.31	4.45	4.93	3.75
50	16.59	14.96	12.22	6.73	12.63	8.24	11.72	7.41
55	19.42	17.48	14.29	8.05	14.71	9.61	13.65	8.48
60	22.34	19.95	16.27	8.93	16.84	11.04	15.63	9.64
65	25.35	22.58	18.36	10.06	19.02	13.01	17.65	11.16
70	28.41	26.04	20.48	10.67	21.22	15.29	19.69	12.43
75	31.59	30.33	22.67	11.56	23.49	19.15	21.80	14.10
76	32.30	31.17	23.15	15.89	24.00	17.15	22.27	14.68

Table 10.3 Comparison of evaluation criteria for the passive and semi-active TLCD systems

RMS Responses				Peak Responses			
Criteria	ATMD	Passive TLCD	Semi-Active TLCD	Criteria	ATMD	Passive TLCD	Semi-Active TLCD
J_1	0.369	0.565	0.449	J_7	0.381	0.631	0.465
J_2	0.417	0.527	0.433	J_8	0.432	0.575	0.483
J_3	0.578	0.641	0.577	J_9	0.717	0.743	0.690
J_4	0.580	0.642	0.579	J_{10}	0.725	0.751	0.700

The results for the evaluation criteria J_1 to J_4 and J_7 to J_{10} for the sample ATMD and the passive and semi-active TLCD systems are presented in Table 10.3. The results of Table 10.3 indicate that the sample ATMD system produces better results for criteria J_1, J_2, J_7, and J_8. In contrast, the semi-active TLCD system outperforms the sample ATMD system for criteria J_3, J_4, J_9, and J_{10}. As such, it is concluded that the performance of the semi-active TLCD system is roughly comparable to that of the ATMD system.

10. 4. Hybrid damper-TLCD system

Agrawal and Yang (1999) present the use of passive dampers for control of the 76-story benchmark building. In their study, a unit damper with capacity of 3.2 x 105 N-sec/m is used, and the optimal distribution of dampers is determined using the Sequential Search Algorithm (SSA) (Zhang and Soong, 1992) and a so-called constrained linear quadratic regulator method. They report that both SSA and the constrained linear quadratic regulator method produce comparable control results. In the SSA method, additional dampers are added and their locations are determined until the responses of a selected floor, the 75th floor, in the benchmark example are smaller than predefined values. The responses of the sample ATMD system are used for the predefined values, and the wind loads applied to the building are stochastic signals defined by the Davenport's cross-power spectral density matrix (Yang et al., 1998). They conclude that the passive damper system can achieve the same level of performance as the sample ATMD system.

For the numerical simulation of the hybrid damper-TLCD system, the same parameters used for the semi-active TLCD and described in the previous section are employed. A unit viscous fluid damper with a capacity of 3.2 x 105 N-sec/m is used and the locations of passive dampers are determined using the SSA method. The RMS acceleration of the 75th floor of the sample ATMD system is used as the predefined value for the SSA. The wind loads applied to the building are the same wind tunnel test data described in the previous section. For the proposed hybrid damper-TLCD system the results obtained from the simulation yielded a total of 10 dampers, 4 in the 74th story and 6 in the 75th story. When only passive dampers are used for the control of the structure, a total of 26 dampers are required in the top ten stories to achieve the same level of performance using the same SSA method. It should be noted that the dampers are most effective in the top stories for control of highrise buildings against wind loading. In contrast, for control of structures against seismic loading, the dampers are generally most effective in the bottom stories.

The comparison of displacement frequency responses of the 75th floor for the uncontrolled, semi-active TLCD, and hybrid damper-TLCD systems is shown in Figure 10.5. This figure shows that the hybrid damper-TLCD system reduces the response of the building significantly more than the semi-active TLCD system at every natural frequency of the building (primary structure). The RMS and peak responses of the selected floors of the benchmark building for the hybrid damper-TLCD system are presented in Table 10.4. The corresponding results for the semi-active TLCD system can be found in the last two columns of Tables 10.1 and 10.2. The results of the evaluation criteria for the hybrid damper-TLCD system are presented in Table 10.5. The corresponding results for the semi-active TLCD system can be found in the fourth and eighth columns of Table 10.3. A comparison of results presented in Table 10.5 and the second and sixth columns of Table 10.3 indicates that the hybrid damper-TLCD system outperforms the sample ATMD system in all criteria except J_7. The performance improvements for the seven criteria J_1 to J_4 and J_8 to J_{10} are 9%, 19%, 10%, 10%, 1%, 8% and 8%, respectively. The value of the criterion J_7 for hybrid damper-TLCD system is 7% more than that for the sample ATMD system.

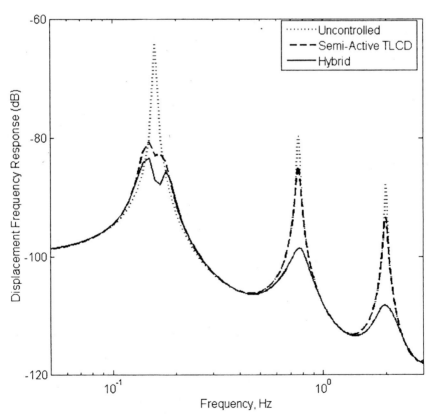

Figure 10.5 Displacement frequency responses of the 75th floor for the uncontrolled, semi-active TLCD, and hybrid damper-TLCD systems

Table 10.4 RMS and peak displacement and acceleration responses of the 76-story building for the hybrid damper-TLCD system

Floor No.	RMS responses		Peak responses	
	Disp. (cm)	Acc. (cm/s^2)	Disp. (cm)	Acc. (cm/s^2)
1	0.01	0.06	0.04	0.21
30	1.14	0.69	4.75	2.91
50	2.75	1.63	11.26	6.17
55	3.21	1.90	13.10	8.05
60	3.69	2.16	14.99	8.73
65	4.17	2.49	16.90	9.96
70	4.66	2.77	18.84	11.05
75	5.17	3.07	20.84	12.35
76	5.28	3.53	21.29	12.70

Table 10.5 Evaluation criteria for the hybrid damper-TLCD system

J_1	J_2	J_3	J_4	J_7	J_8	J_9	J_{10}
0.336	0.339	0.521	0.524	0.407	0.431	0.659	0.668

Figure 10.6 shows the displacement response time histories of the 75th floor for both the uncontrolled structure and the hybrid damper-TLCD system. Figure 10.7 shows the corresponding results for the acceleration response time histories. These figures as well as Tables 10.2 and 10.4 demonstrate clearly that the hybrid damper-TLCD system reduces both displacements and accelerations significantly compared with the uncontrolled structure.

Two additional numerical simulations are carried out to evaluate the robustness of the proposed hybrid damper-TLCD system and its sensitivity to modeling errors using the same 76-story building benchmark control problem. The sensitivity analysis is performed in terms of the stiffness of the structure. In one simulation the stiffness of the structure is increased by 15% ($\Delta K = +15\%$) and in another simulation it is decreased by 15% ($\Delta K = -15\%$), as suggested by Yang et al. (2000). The controller configuration obtained for the building with the previous value of stiffness is applied to the building with $\Delta K = \pm 15\%$ and the same time-history response analyses are carried out.

The resulting RMS and peak displacement and acceleration responses of the selected floors of the benchmark building for the hybrid damper-TLCD systems are presented in Tables 10.6 and 10.7, respectively. Table 10.8 presents the results for the evaluation criteria. As observed from the results in Tables 10.6-10.8, the hybrid damper-TLCD system is robust in terms of the stiffness modeling error for the control of both displacement and acceleration responses. In particular, the values of RMS and peak

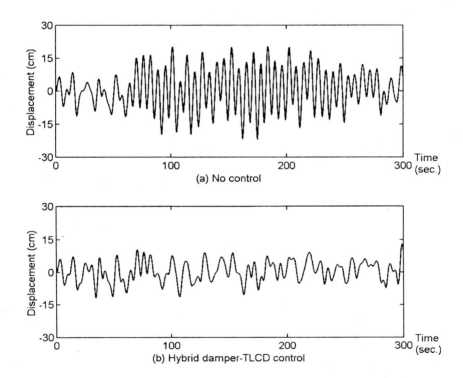

Figure 10.6 Displacement response time histories of the 75th floor subjected to the wind tunnel test loads

accelerations of the 75th floor as well as the top floor (76th floor) of the building with $\Delta K = \pm 15\%$ are all within the allowable maximum values for the floor accelerations, 5 cm/s² and 15 cm/s², respectively, as mentioned earlier.

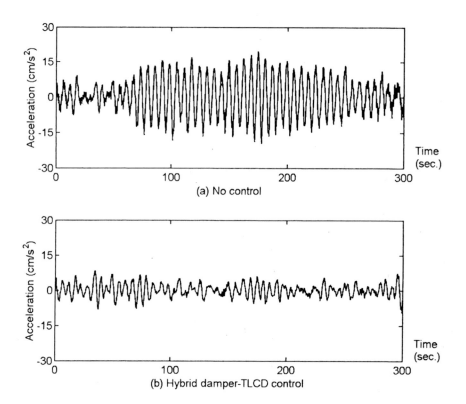

(a) No control

(b) Hybrid damper-TLCD control

Figure 10.7 Acceleration response time histories of the 75th floor subjected to the wind tunnel test loads

Table 10.6 RMS displacement and acceleration responses of the 76-story building for the hybrid control system using the building with uncertainty in stiffness matrix

Floor No.	$\Delta K = +15\%$		$\Delta K = -15\%$	
	Disp. (cm)	Acc. (cm/s^2)	Disp. (cm)	Acc. (cm/s^2)
1	0.01	0.06	0.01	0.06
30	1.00	0.73	1.43	0.75
50	2.41	1.74	3.44	1.78
55	2.81	2.02	4.02	2.09
60	3.22	2.31	4.61	2.38
65	3.65	2.66	5.22	2.72
70	4.08	2.96	5.84	3.03
75	4.52	3.34	6.48	3.73
76	4.62	3.42	6.63	3.35

Table 10.7 Peak displacement and acceleration responses of the 76-story building for the hybrid control system using the building with uncertainty in stiffness matrix

Floor No.	$\Delta K = +15\%$		$\Delta K = -15\%$	
	Disp. (cm)	Acc. (cm/s^2)	Disp. (cm)	Acc. (cm/s^2)
1	0.04	0.20	0.04	0.22
30	4.50	2.79	5.26	2.76
50	10.68	6.98	12.47	6.78
55	12.43	7.95	14.51	8.39
60	14.22	8.57	16.60	9.39
65	16.05	10.02	18.73	11.48
70	17.89	10.89	20.89	12.32
75	19.79	12.44	23.11	13.99
76	20.22	12.16	23.61	13.87

Table 10.8 Evaluation criteria of the hybrid control system for the 76-story building with uncertainty in stiffness matrix

RMS Responses			Peak Responses		
Criteria	$\Delta K = +15\%$	$\Delta K = -15\%$	Criteria	$\Delta K = +15\%$	$\Delta K = -15\%$
J_1	0.365	0.408	J_7	0.410	0.461
J_2	0.362	0.378	J_8	0.437	0.474
J_3	0.456	0.654	J_9	0.626	0.731
J_4	0.458	0.656	J_{10}	0.634	0.740

10. 5. Stochastic wind loads

To evaluate the effectiveness of the proposed hybrid damper-TLCD control system under various types of wind loads, $F(t)$ in Eqs. (10.9) and (10.10), the stochastic wind loads defined by the Davenport's cross-power spectral density matrix are applied to the benchmark building. The (i, j) element of Davenport's cross-power spectral density matrix defined in the frequency domain, $S_{ww}(\omega)$, is expressed as (Yang et al., 1998)

$$S_{w_i w_j}(\omega) = \frac{8\overline{w}_i \overline{w}_j K_0 V_r^2}{\overline{V}_i \overline{V}_j |\omega|} \frac{(600\omega / \pi V_r)^2}{\left[1 + (600\omega / \pi V_r)^2\right]^{4/3}} \exp\left(-\frac{c_1 |\omega|}{2\pi} \frac{|h_i - h_j|}{V_r}\right)$$

(10.12)

where ω is the frequency in radians per second, $\overline{w}_i$ is the average wind force on the ith floor, $\overline{V}_i$ is the mean wind velocity at the ith floor, V_r is the reference mean wind velocity in meters per second at 10 m above the ground, h_i is the height of the ith floor, K_0 is a constant depending on the surface roughness of the ground, and c_1 is a constant which depends on different factors such as terrain roughness, height above ground, and wind speed. Simiu and Scanlan (1996) present empirical values for this coefficient in terms of the mean wind speed at 10 m above the ground in the range 2 to 10.

Following Yang et al. (1998), the values of $K_0 = 0.03$, $c_1 = 7.7$, and $V_r = 15$ m/sec are used for the parameters in this study. The first 5 minutes of time histories of resulting wind loads on the 66th and 70th floors are displayed in Figure 10.8.

For the numerical simulation of the hybrid damper-TLCD system using the stochastic wind loads, the same controller configuration obtained previously using the wind tunnel test loads is applied. Figure 10.9 shows the displacement response time histories of the 75th floor for both uncontrolled structure and the hybrid damper-TLCD system subjected to the stochastic wind loads. Figure 10.10 shows the corresponding results for the acceleration

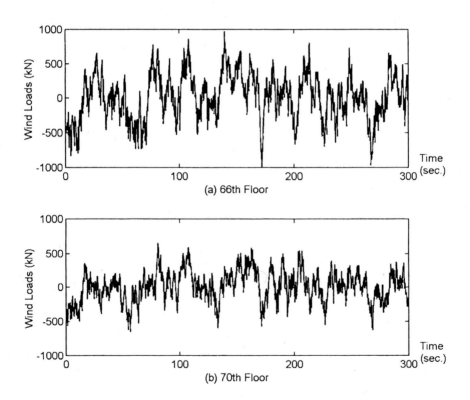

Figure 10.8 Time histories of stochastic wind loads acting on the 66th and 70th floors

response time histories. These figures again show clearly the hybrid damper-TLCD system reduces both displacements and accelerations significantly compared with the uncontrolled structure under the wind loading.

10. 6.   Concluding remarks

The effectiveness of both a semi-active TLCD system and the hybrid damper-TLCD control system was investigated for control of wind-induced motion of a 76-story benchmark building. It is shown that the semi-active TLCD control system performs comparably to a sample ATMD system. Considering the fact that the semi-active TLCD system does not need any actuator requiring a large electro-mechanic capacity and thus is able to

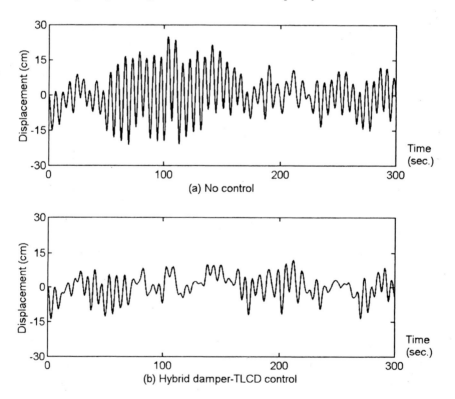

Figure 10.9 Displacement response time histories of the 75th floor subjected to the stochastic wind loads

operate with only small power, such as a battery, it is concluded that the semi-active TLCD system is an attractive alternative to the ATMD system. Further, the TLCD system provides several advantages over the TMD system as described in section 2.2.4.

By judiciously integrating the semi-active TLCD system with a passive supplementary damper system, the hybrid damper-TLCD system provides reliable and robust control of wind-induced vibrations of highrise buildings in terms of power or computer failure. It is shown that the hybrid system can reduce the response of the building significantly more than the semi-active TLCD system at every natural frequency of the building. Moreover, the

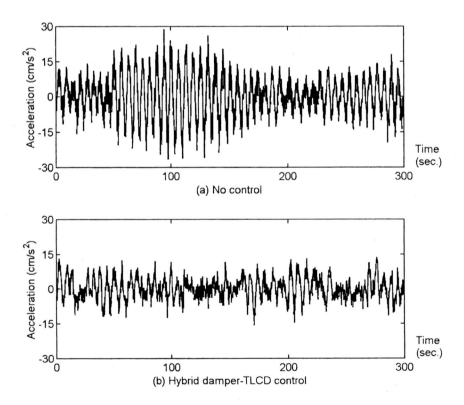

Figure 10.10 Acceleration response time histories of the 75th floor subjected to the stochastic wind loads

hybrid damper-TLCD system is robust in terms of the stiffness modeling error for the control of both displacement and acceleration responses. Furthermore, the simulation results using stochastic wind loads clearly show that the proposed hybrid control system can perform effectively under various wind loads.

Chapter 11

Vibration Control of Cable-Stayed Bridges

11. 1. Introduction

Cable-stayed bridges have recently gained increasing popularity due to their economic and aesthetic advantages. These bridges, however, are flexible and control of their vibrations is an important consideration and a challenging problem. It has been recognized that the analysis and control of cable-stayed bridges is a challenging problem with complexities in structural modeling, and control design and implementation. In particular, the geometric nonlinearity of cables due to the change of shape under varying stresses makes the analysis of cable-stayed bridges more complicated compared with other types of bridge structures (Adeli and Zhang, 1995).

Articles on vibration control of cable-stayed bridges have appeared in the literature recently. Cable stays in such bridges provide relatively small intrinsic damping. Therefore, early studies on control of cable-stayed bridge have concentrated on increasing the damping capacity of bridge using passive supplementary dampers. Ali and Abdel-Ghaffar (1994) discuss the effectiveness, feasibility, and limitations of passive supplementary dampers for cable-stayed bridges analytically. They conclude that when passive damper devices are installed at critical zones, such as between deck and abutment and between deck and tower, the inelastic behavior of cable-stayed bridges can be minimized. A similar experimental study using passive damper devices for control of cable-stayed bridges subjected to seismic loads has been reported by Villaverde and Marin (1995). Tabatabai and Mehrabi (2000) discuss design of mechanical viscous dampers for passive control of cable-stayed bridges under wind induced or galloping vibrations. They present simplified formulations based on the fundamental mode of vibrations for finding the capacities of dampers and their location on the stays.

In addition to passive supplementary dampers, a number of studies on active and semi-active control of cable-stayed bridges have also been reported in the literature. Warnitchai et al. (1993) investigate experimentally active tendon control of cable-stayed bridges subjected to a vertical sinusoidal force. Experiments were performed using a simple cable-supported cantilever beam. Schemmann and Smith (1998a and b) investigate the effectiveness of active control of cable-stayed bridges using an LQR feedback control algorithm and discuss issues involved such as geometric nonlinearity and high-order vibration modes. They conclude that control of higher-order modes is critical and actuators located close to the center of bridge span are the most effective for control of the structural response. Bossens and Preumont (2001) present a scheme for active tendon control of cable-stayed bridges subjected to wind and earthquake loading using collocated actuator/sensor pairs and verify it with experimental results on scaled models. He et al. (2001) present semi-active control of a cable-stayed bridge using resetting semi-active stiffness dampers. They show that semi-active control of the bridge reduces the response more significantly than passive supplementary dampers.

In this chapter, the wavelet-hybrid feedback LMS algorithm presented in section 8.3 is used for vibration control of cable-stayed bridges under various seismic excitations. Its effectiveness is investigated through numerical simulation using the benchmark control problem created by Dyke et al. (2000) and described briefly in the next section. The performance of the new algorithm is compared with that of a sample LQG controller. Additional numerical simulations are performed to evaluate the sensitivity of the control model to modeling errors and verify its robustness.

11. 2.   Cable-stayed bridge benchmark problem

The benchmark control problem used for simulation in this study is based on the Bill Emerson Memorial Bridge in Cape Girardeau, Missouri, USA. The bridge spans the Mississippi river and connects the states of Missouri and Illinois. It consists of a semi-fan type cable-stayed bridge with two main concrete towers and a deck which extends over 12 additional piers in the approach bridge from the Illinois side. In the benchmark control problem, only the cable-stayed part of bridge is used as shown in Figure 11.1.

In the cable-stayed part of the bridge, the main span has a length of 350.6 m and each side span has a length of 142.7 m. The heights of H-shaped towers are 100 m at pier II and 105 m at pier III (Figure 11.1). A total of 128 cables, made of high-strength, low-relaxation steel, are evenly supported by two towers, that is, 64 cables on each tower. The deck of width 29.3 m is built with prestressed concrete slabs and steel beams.

Dyke et al. (2000) present a benchmark control problem for the cable-

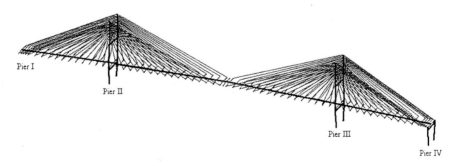

Figure 11.1 3-D view of the benchmark cable-stayed bridge

stayed bridge. A three-dimensional finite element model of the bridge is created using ABAQUS (1998). Two-node shear beam elements are used to model the beams and two-node linear space truss elements are used to model the cables. Geometric nonlinearity due to cable sag effect is taken into account approximately using an equivalent modulus of elasticity (Adeli and Zhang, 1995). The resulting model has 419 degrees of freedom. The first six natural frequencies of the structure are 0.2889 Hz, 0.3699 Hz, 0.4683 Hz, 0.5158 Hz, 0.5812 Hz, and 0.6490 Hz.

Eighteen criteria, denoted by J_1 to J_{18}, are provided to evaluate the control performance. The first six performance criteria (J_1 to J_6) are non-dimensional ratios of the responses of the controlled bridge to those of the uncontrolled bridge subjected to three earthquake records, the El Centro (California, 1940), Mexico City (Mexico, 1985), and Gebze (Turkey, 1999) earthquakes, shown in Figure 11.2. They are the maximum values of the base shear and shear at the deck level in the two towers (J_1 and J_2, respectively), the maximum values of the base moment and moment at the deck level in the two towers (J_3 and J_4, respectively), the maximum cable sag or deviation (J_5), and the maximum deck displacement (J_6).

The next five performance criteria (J_7 to J_{11}) are non-dimensional ratios of the normed responses of the controlled bridge to those of the uncontrolled bridge subjected to the same three earthquakes, where the normed value of a response is defined as

$$\|\bullet\| = \sqrt{\frac{1}{t_f} \int_0^{t_f} (\bullet)^2 \, dt} \tag{11.1}$$

in which t_f is the time required for the response to attenuate. They are the maximum normed values of the base shear and shear at the deck level in the towers (J_7 and J_8, respectively), the maximum normed values of the base

moment and moment at the deck level in the towers (J_9 and J_{10}, respectively), and the maximum normed value of the cable sag or deviation (J_{11}).
The next four performance criteria (J_{12} to J_{15}) are non-dimensional measures

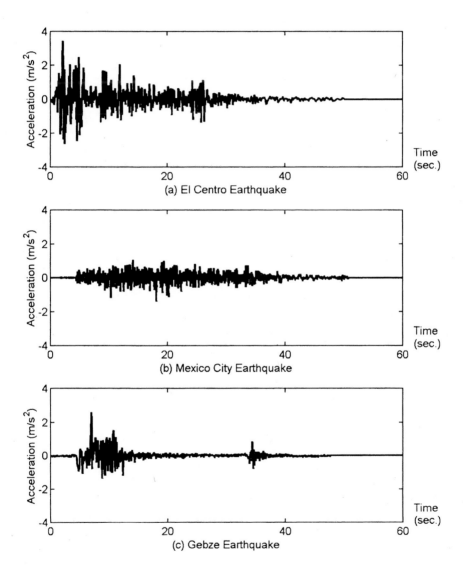

Figure 11.2 Time histories of the El Centro, Mexico City, and Gebze earthquake acceleration records

of the control device performances. They are the maximum force generated by all the control devices normalized by the weight of the bridge superstructure (J_{12}), the maximum stroke of all the control devices normalized by the maximum uncontrolled displacement at the top of the two towers relative to the ground (J_{13}), the maximum instantaneous power required to control the bridge normalized by the product of the peak uncontrolled velocity at the top of the two towers relative to ground and the weight of the bridge superstructure (J_{14}) where the instantaneous power is given by the absolute value of the product of the velocity and the force generated by the control device, and the integration of instantaneous power over time normalized by the product of the weight of the bridge superstructure and the maximum uncontrolled displacement at the top of the two towers relative to the ground (J_{15}).

The last three performance criteria are the total number of control devices (J_{16}), the total number of sensors (J_{17}), and the dimension of the discrete time state vector required to implement the control algorithm (J_{18}).

11. 3. Numerical simulation

For the sake of comparison, the same numbers of devices and sensors are used for both the new and the sample LQG control algorithms. Also, the same number is used for the dimension of the discrete time state vector required to implement the control algorithm (the last three rows in Table 11.1). Numerical simulation results are displayed in Figures 11.3 to 11.8. Figures 11.3, 11.5, and 11.7 show the uncontrolled and controlled time histories of base shear force and base moment at pier II subjected to El Centro, Mexico City, and Gebze earthquake records, respectively. Figures 11.4, 11.6, and 11.8 show the time histories of base shear force at pier II subjected to El Centro, Mexico City, and Gebze earthquake records, respectively, using the sample LQG and the wavelet-hybrid feedback-LMS control algorithms. It is clear from the results that the responses of the cable-stayed bridge can be significantly reduced by using the wavelet-hybrid feedback-LMS control algorithm. Results also show that the new control algorithm is more effective than the sample LQG controller for all three earthquake records.

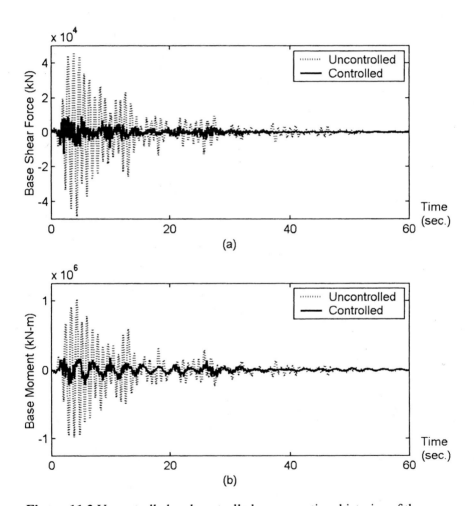

Figure 11.3 Uncontrolled and controlled response time histories of the benchmark bridge subjected to El Centro earthquake record: a) base shear force at pier II, b) base moment at pier II

The results of evaluation criteria are summarized in Table 11.1. The maximum base shear (J_1) in towers using the wavelet-hybrid feedback-LMS control model is 16%, 13%, and 6% less than the corresponding values using the sample LQG model when the bridge is subjected to El Centro, Mexico City, and Gebze earthquake records, respectively. The corresponding normed values J_7 are 13%, 24%, and 8% less than the corresponding values using the sample LQG model, respectively. The maximum shear at deck level (J_2) in towers using the wavelet-hybrid feedback-LMS control model is 7%, 17%, and 11% less than the corresponding values using the sample LQG model when the bridge is subjected to El Centro, Mexico City, and Gebze earthquake records, respectively. The corresponding normed values J_8 are 23%, 25%, and 15% less than the corresponding values using the sample

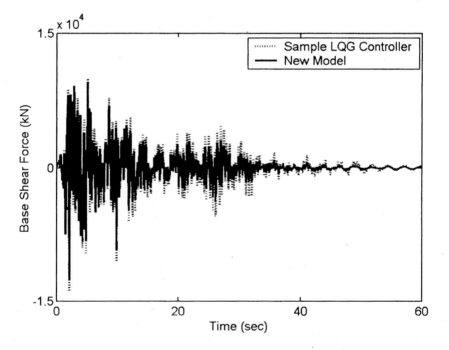

Figure 11.4 Time histories of base shear force at pier II of the benchmark bridge subjected to El Centro earthquake record using the sample LQG and the wavelet-hybrid feedback-LMS control algorithms

LQG model, respectively. Similar performance improvements are noted for evaluation criteria J_3 to J_6 and J_9 to J_{11}.

In terms of the required control forces and power, criteria J_{12} to J_{15}, the values for the new model are 3-42% greater than those for the sample LQG control algorithm.

Table 11.1 Comparison of evaluation criteria

Criteria	Sample LQG Controller			WHFL Controller		
	El Centro	Mexico	Gebze	El Centro	Mexico	Gebze
J_1	0.3970	0.4969	0.4594	0.3344	0.4287	0.4332
J_2	1.0696	1.2706	1.3775	0.9922	1.0537	1.2300
J_3	0.2943	0.5858	0.4413	0.2742	0.4535	0.3894
J_4	0.6455	0.6820	1.2234	0.6484	0.4808	0.9533
J_5	0.1825	0.0770	0.1501	0.1706	0.0679	0.1238
J_6	1.2033	2.3938	3.6042	1.1173	1.6588	2.3374
J_7	0.2353	0.4554	0.3359	0.2052	0.3481	0.3099
J_8	1.2018	1.2566	1.4822	0.9308	0.9433	1.2532
J_9	0.2703	0.4551	0.4633	0.2229	0.3503	0.3919
J_{10}	0.8922	1.1251	1.4730	0.6581	0.7815	1.0184
J_{11}	2.830E-02	1.043E-02	1.725E-02	2.384E-02	8.939E-03	1.399E-02
J_{12}	1.961E-03	6.243E-04	1.831E-03	2.161E-03	8.717E-04	1.961E-03
J_{13}	5.834E-06	8.402E-06	2.726E-05	6.417E-06	8.855E-06	2.833E-05
J_{14}	3.003E-02	1.043E-02	3.477E-02	3.381E-02	1.476E-02	3.564E-02
J_{15}	4.435E-09	1.454E-09	9.594E-09	4.993E-09	2.057E-09	9.835E-09
J_{16}	24	24	24	24	24	24
J_{17}	9	9	9	9	9	9
J_{18}	30	30	30	30	30	30

11. 4. Sensitivity analysis

Additional numerical simulations are carried out to evaluate the robustness of the new wavelet-hybrid feedback-LMS algorithm and its sensitivity to modeling errors. Due to geometric nonlinearity, the stiffness of the cable-stayed bridge may change during strong ground motions. Further, the dynamic characteristics of the finite element model may not be identical to those of the real bridge. For the purpose of the sensitivity analysis the

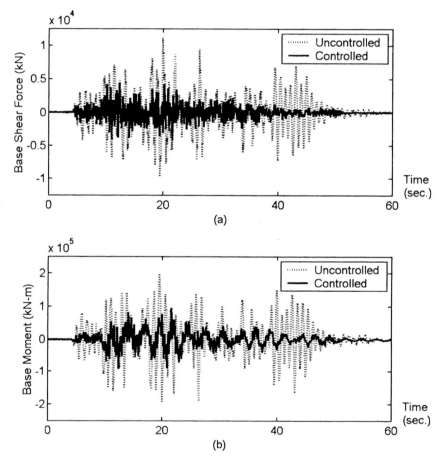

Figure 11.5 Uncontrolled and controlled response time histories of the benchmark bridge subjected to Mexico City earthquake record: a) base shear force at pier II, b) base moment at pier II

following perturbation is introduced in the structural stiffness matrix:

$$K_{pert} = K(1+\Delta) \tag{11.2}$$

where Δ is the perturbation ratio, and K_{pert} is the resulting perturbed structural stiffness matrix.

When the stiffness of the actual structure is greater than that used in the mathematical finite element model, simulation results show no adverse effect on the vibration control of the cable-stayed bridge, as expected. On the other hand, when the stiffness of the actual structure is smaller than that used in the mathematical finite element model, simulation results show deterioration in the control performance. Simulations were performed by gradually increasing the magnitude of the perturbation ratio, Δ, in order to investigate

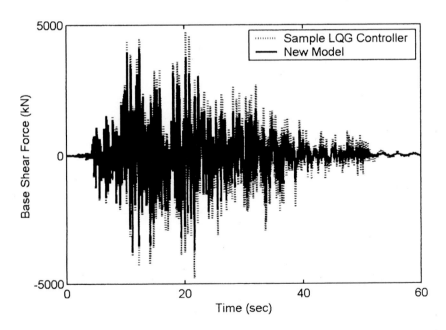

Figure 11.6 Time histories of base shear force at pier II of the benchmark bridge subjected to Mexico City earthquake record using the sample LQG and the wavelet-hybrid feedback-LMS control algorithms

the stability of the new control algorithm. Significant vibration reduction and stable results were obtained with values of perturbation ratio up to -0.07. When Δ = -0.07 the control deterioration is 0.5-12% for the El Centro earthquake, 7-40% for the Mexico City earthquake, and 2-23% for the Gebze earthquake.

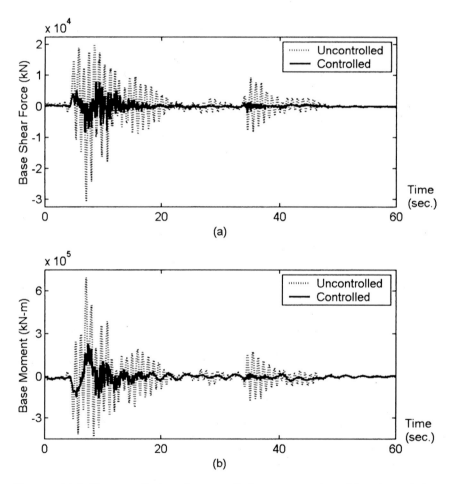

Figure 11.7 Uncontrolled and controlled response time histories of the benchmark bridge subjected to Gebze earthquake record: a) base shear force at pier II, b) base moment at pier II

Consequently, the stability threshold for the perturbation ratio is found to be around -0.07. A similar observation is reported in Turan et al. (2002) where the μ-synthesis feedback control method is used for control of the benchmark cable-stayed bridge.

Figures 11.9 to 11.11 show the uncontrolled and controlled time histories of base shear force and base moment at pier II subjected to three earthquakes used previously when Δ = -0.07. Table 11.2 summarizes the results of evaluation criteria for all three earthquakes. It is observed from the results that no major performance difference in the perturbed system occurs, thereby proving the robustness of the new control algorithm.

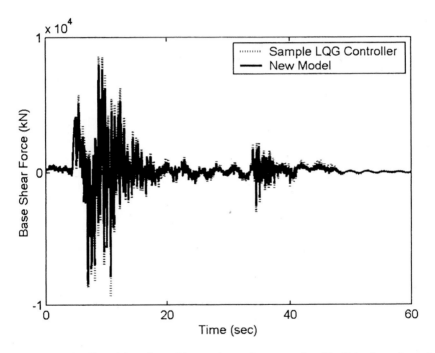

Figure 11.8 Time histories of base shear force at pier II of the benchmark bridge subjected to Gebze earthquake record using the sample LQG and the wavelet-hybrid feedback-LMS control algorithms

Table 11.2 Evaluation criteria results for $\Delta = -0.07$

Criteria	El Centro	Mexico	Gebze
J_1	0.3583	0.6003	0.4594
J_2	1.1167	1.3161	1.2689
J_3	0.2784	0.5395	0.4104
J_4	0.6634	0.5437	0.9711
J_5	0.1712	0.0725	0.1527
J_6	1.0759	1.9653	2.4250
J_7	0.2271	0.4545	0.3821
J_8	1.1608	1.3495	1.8646
J_9	0.2437	0.4384	0.4570
J_{10}	0.7039	0.8967	1.1395
J_{11}	2.476E-02	9.494E-03	1.672E-02
J_{12}	2.161E-03	1.020E-03	1.961E-03
J_{13}	8.216E-06	9.205E-06	2.911E-05
J_{14}	3.114E-02	1.377E-02	3.621E-02
J_{15}	4.599E-09	1.920E-09	9.991E-09
J_{16}	24	24	24
J_{17}	9	9	9
J_{18}	30	30	30

11. 5. Concluding remarks

In this chapter, the wavelet-hybrid feedback-LMS control algorithm was used for vibration control of cable-stayed bridges under various seismic excitations. To evaluate the performance, simulations are performed on a cable-stayed bridge benchmark control problem. Simulation results demonstrate that the new control algorithm is more effective than the sample LQG controller for all three earthquake records consistently.

Moreover, the results of the sensitivity analysis show that the algorithm is stable even when the structural stiffnesses are underestimated by a relatively large value of 7%. This number should be considered in the context of nonlinear behavior of cable-stayed bridges. For control of highrise building structures subjected to wind loading, results provided in Chapter 10 indicate that the control algorithm produces stable results for a much larger value of the perturbation ratio. Consequently, it is concluded that the new control algorithm is robust against the uncertainties existing in modeling structures.

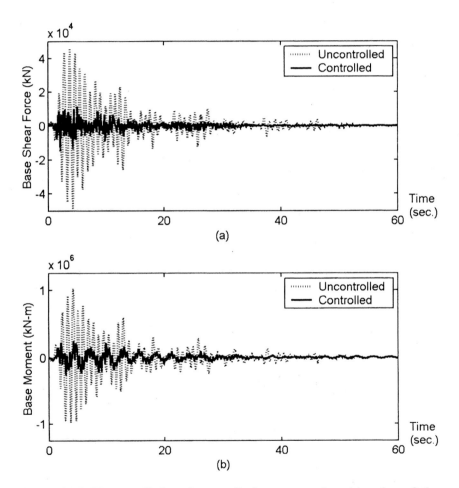

Figure 11.9 Uncontrolled and controlled response time histories of the benchmark bridge with stiffness perturbation Δ = -0.07 subjected to El Centro earthquake record: a) base shear force at pier II, b) base moment at pier II

188

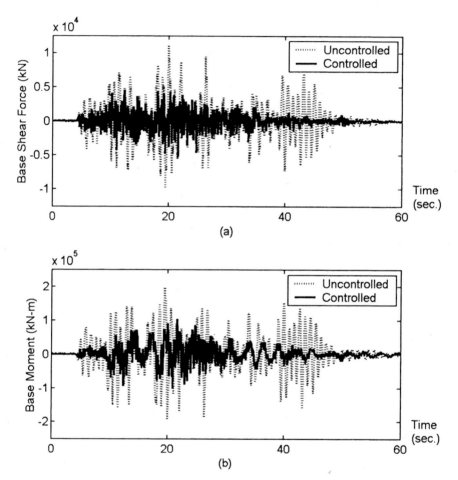

Figure 11.10 Uncontrolled and controlled response time histories of the benchmark bridge with stiffness perturbation $\Delta = -0.07$ subjected to Mexico City earthquake record: a) base shear force at pier II, b) base moment at pier II

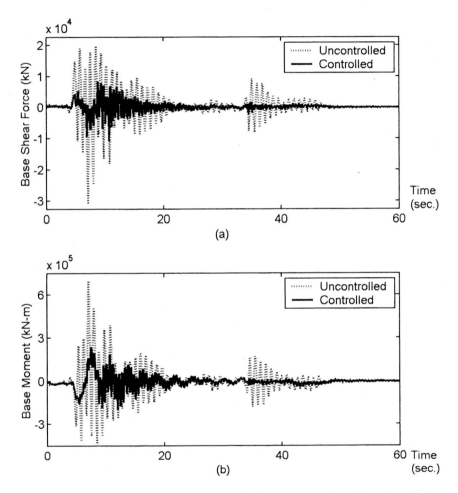

Figure 11.11 Uncontrolled and controlled response time histories of the benchmark bridge with stiffness perturbation $\Delta = -0.07$ subjected to Gebze earthquake record: a) base shear force at pier II, b) base moment at pier II

Chapter 12

Conclusion – Toward a New Generation of Smart Building and Bridge Structures

A new control algorithm, wavelet-hybrid feedback-LMS algorithm, is developed to overcome the shortcomings of the classical feedback control algorithms and the filtered-x LMS control algorithm. The new control algorithm integrates a feedback control algorithm such as the LQR or LQG algorithm with the filtered-x LMS algorithm and utilizes a wavelet multi-resolution analysis for the low-pass filtering of external dynamic excitations. Due to the integration, the total control force is obtained by summing the control force determined by the filtered-x LMS controller and the control force obtained through the feedback controller. Simulation results show that since the control forces determined by the filtered-x LMS algorithm are adapted by updating the FIR filter coefficients at each sampling time until the output error is minimized, the combination of a classical feedback controller with a filtered-x LMS controller results in effective control of both steady-state and transient vibrations. Also, it is shown that the new algorithm is capable of suppressing vibrations over a range of input excitation frequencies unlike the classic feedback control algorithms whose control effectiveness decreases considerably when the frequency of the external disturbance differs from the fundamental frequency of the system. Further, the advantage of the proposed algorithm is that the external excitation is included in the formulation.

The higher frequency contents of external excitations such as earthquakes, winds, and ocean waves impede the stabilization of the FIR filter during adaptation. Further, the frequency bandwidths of such environmental signals are much wider than those of common structural systems. Therefore, the use of a low-pass filter that eliminates higher frequency components of the external excitation is crucial in order to apply the algorithm for control of civil structures. This can be effective because the response of most civil structures is not affected by high frequency contents of the external excitations by any significant amount.

A wavelet based low-pass filtering is proposed for stable adaptation of the FIR filter coefficients. Considering the fact that the orthogonal wavelet filtering requires only integer operations, real time control of large structures can be achieved with little additional computational efforts due to filtering. Moreover, the wavelet transform provides an effective way of processing non-stationary signals, to which most environmental signals belong, due to the locality of the basis function of the wavelet in both time and frequency domains. Simulation results demonstrate the wavelet transform can be effectively used as a low-pass filter for control of civil structures without any significant additional computational burden.

A new hybrid control system, the hybrid damper-TLCD system, is proposed, and its performance is evaluated for control of responses of 3D irregular buildings under various seismic excitations and for control of wind-induced motion of highrise buildings. It is developed through judicious integration of a passive supplementary damping system with a semi-active TLCD system.

For both supplementary damper and TLCD systems, damping is

achieved and damping forces are controlled through an orifice/valve, making them suitable not only for passive control systems but also for semi-active control systems. However, it is shown that the performance improvement of semi-active viscous fluid damper systems over the less complicated and less costly passive damper systems is not always guaranteed depending on the flexibility of the structure, while a semi-active TLCD system can reduce the response significantly compared with a passive TLCD system. As such, by integrating a passive supplementary damping system with a semi-active TLCD system, the new hybrid control system utilizes the advantages of both passive and semi-active control systems along with improving the overall performance significantly. Additionally, the proposed hybrid control system eliminates the need for a large power requirement, unlike other proposed hybrid control systems, where active and passive systems are combined.

Simulations performed on irregular 3D building structures and a 76-story building show that the new hybrid control system is effective in significantly reducing the response of structures under seismic excitations as well as wind loads. It is also demonstrated that the hybrid control system provides increased reliability and maximum operability during normal operations as well as a power or computer failure. Further, it is shown that the hybrid damper-TLCD system is robust in terms of the stiffness modeling error for the control of both displacement and acceleration responses.

Finally, the wavelet-hybrid feedback-LMS control algorithm is used for vibration control of cable-stayed bridges under various seismic excitations. To evaluate the performance, simulations are performed on a cable-stayed bridge benchmark control problem. Simulation results demonstrate that the proposed algorithm is effective for control of cable-stayed bridges. Results also show that the new control algorithm is more effective than the sample LQG controller for three different earthquake records consistently. Moreover, the simulation results at which the structural stiffness matrices are perturbed show that the control algorithm is well performing and robust against the uncertainty existing in the modeling of the bridge.

The semi-active TLCD system described in this book requires a controllable orifice/valve. It is assumed that the valve dynamics are negligible and the head loss coefficient of the orifice (or valve opening ratio) can be ideally changed continuously by applying a command signal. Although useful for design purposes, this ideal model may not accurately describe the nonlinear dynamic behavior of the TLCD system. Therefore, further research can include the valve dynamics in the formulation of the control problem.

Further research is recommended to include the response time of the orifice/valve to the command signal in the formulation. It is also recommended that the modeling of the orifice-controlled semi-active TLCD system as well as the effectiveness of the new control algorithm be verified by experiments.

A study on the response time and orifice dynamics of the semi-active magnetorheological (MR) damper, which also requires a controllable orifice, is performed analytically and experimentally by Yang et al. (2001). They also suggest that the response time and orifice dynamics of the semi-active device be included in the control formulation for more accurate design of the control system and show that the pulse width modulation (PWM)-based current driver can be effective in reducing the response time of the MR damper. Since both semi-active TLCD and semi-active MR damper systems utilize similar controllable orifices, the study on the MR damper can be extended to that of the semi-active TLCD system.

References

ABAQUS (1998), Hibbitt, Karlsson & Soresen Inc. Pawtucket, RI.

Abe, M. (1996), "Semi-Active Tuned Mass Dampers for Seismic Protection of Civil Structures," *Earthquake Engineering & Structural Engineering*, Vol. 25, No. 7, pp. 743-749.

Abry, P. (1997), "Ondelettes et turbulence". *Multirésolutions, Algorithmes de Décomposition, Invariance D'échelles,* Diderot Editeur, Paris, France.

Achkire, Y. and Preumont, A. (1996), "Active Tendon Control of Cable-Stayed Bridges," *Earthquake Engineering & Structural Dynamics*, Vol. 25, No. 6, pp. 585-597.

Adeli, H. (2001), "Neural Networks in Civil Engineering: 1989–2000," *Computer-Aided Civil and Infrastructure Engineering*, Vol. 16, No. 2, pp. 126-142.

Adeli, H. and Ghosh-Dastidar, S. (2004), "Mesoscopic-Wavelet Freeway Work Zone Flow and Congestion Feature Extraction Model," *Journal of Transportation Engineering*, ASCE, Vol. 130, No. 1, pp. 94-103.

Adeli, H., Ghosh-Dastidar, S., and Dadmehr, N. (2007), "A Wavelet-Chaos Methodology for Analysis of EEGs and EEG Sub-bands to detect Seizure and Epilepsy," *IEEE Transactions on Biomedical Engineering*, Vol. 54, No. 2, pp. 205-211.

Adeli, H. and Hung, S. L. (1993a), "A Concurrent Adaptive Conjugate Gradient Learning Algorithm on MIMD Machines," *Journal of Supercomputer Applications*, MIT Press, Vol. 7, No. 2, pp. 155-166.

Adeli, H. and Hung, S. L. (1993b), "A Fuzzy Neural Network Learning Model for Image Recognition," *Integrated Computer-Aided Engineering*, Vol. 1, No. 1, pp. 43-55.

Adeli, H. and Hung, S. L. (1994), "An Adaptive Conjugate Gradient Learning Algorithm for Effective Training of Multilayer Neural Networks," *Applied Mathematics and Computation*, Vol. 62, No. 1, pp. 81-102.

Adeli, H. and Hung, S. L. (1995), *Machine Learning — Neural Networks, Genetic Algorithms, and Fuzzy Systems*, John Wiley, New York.

Adeli, H. and Jiang, X. (2006), "Dynamic Fuzzy Wavelet Neural Network Model for Structural System Identification," *Journal of Structural Engineering*, ASCE, Vol. 132, No. 1, pp. 102-111.

Adeli, H. and Karim, A. (2000), "A Fuzzy-Wavelet RBF Neural Network Model for Freeway Incident Detection," *Journal of Transportation Engineering*, ASCE, Vol. 126, No. 6, pp. 464-471.

Adeli, H. and Karim, A. (2001), *Construction Scheduling, Cost Optimization, and Management — A New Model Based on Neurocomputing and Object Technologies*, Spon Press, London.

Adeli, H. and Karim, A. (2005), *Wavelets in Intelligent Transportation Systems*, John Wiley and Sons, New York.

Adeli, H. and Kim, H. (2004), "Wavelet-Hybrid Feedback Least Mean Square Algorithm for Robust Control of Structures," *Journal of Structural Engineering*, ASCE, Vol. 130, No. 1, pp. 128-137.

Adeli, H. and Park, H. S. (1995a), "Counter Propagation Neural Network in Structural Engineering," *Journal of Structural Engineering*, ASCE, Vol. 121, No. 8, pp. 1205-1212.

Adeli, H. and Park, H. S. (1995b), "A Neural Dynamics Model for Structural Optimization — Theory," *Computers and Structures*, Vol. 57, No. 3, 1995, pp. 383-390.

Adeli, H. and Park, H. S. (1998), *Neurocomputing for Design Automation*, CRC Press, Boca Raton, Florida.

Adeli, H. and Saleh, A. (1997), "Optimal Control of Adaptive/Smart Bridge Structures," *Journal of Structural Engineering*, ASCE, Vol. 123, No. 2, pp. 218-226.

Adeli, H. and Saleh, A. (1998), "Integrated Structural/Control Optimization of Large Adaptive/Smart Structures," *International Journal of Solids and Structures*, Vol. 35, Nos. 28-29, pp. 3815-3830.

Adeli, H. and Saleh, A. (1999), *Control, Optimization, and Smart Structures — High-Performance Bridges and Buildings of the Future*, John Wiley and Sons, New York.

Adeli, H. and Samant, A. (2000), "An Adaptive Conjugate Gradient Neural Network — Wavelet Model for Traffic Incident Detection," *Computer-Aided Civil and Infrastructure Engineering*, Vol. 13, No. 4, pp. 251-260.

Adeli, H. and Yeh, C. (1989), "Perceptron Learning in Engineering Design," *Microcomputers in Civil Engineering*, Vol. 4, No. 4, pp. 247-256.

Adeli, H. and Zhang, J. (1993), "An Improved Perceptron Learning Algorithm," *Neural, Parallel, and Scientific Computations*, Vol. 1, No. 2, pp. 141-152.

Adeli, H. and Zhang, J. (1995), "Fully Nonlinear Analysis of Composite Girder Cable-Stayed Bridges," *Computers and Structures*, Vol. 54, No. 2, pp. 267-277.

Adeli, H., Ghosh-Dastidar, S., and Dadmehr, N. (2007), "A Wavelet-Chaos Methodology for Analysis of EEGs and EEG Sub-bands to Detect Seizure and Epilepsy," *IEEE Transactions on Biomedical Engineering*, Vol. 54, No. 2, February, pp. 205-211.

Adeli, H., Zhou, Z., and Dadmehr, N. (2003), "Analysis of EEG Records in an Epileptic Patient Using Wavelet Transform," *Journal of Neuroscience Methods*, Vol. 123, No. 1, pp. 69-87.

Agrawal, A. K. and Yang, J. N. (1999), "Passive Damper Control of the 76-Story Wind-Excited Benchmark Building," *Proceedings of the Second World Conference on Structural Control*, Chichester, England; New York, Vol. 2, pp. 1481-1490.

AISC (1998), *Manual of Steel Construction — Load and Resistance Factor Design, Vol. I, Structural Members, Specifications, and Codes*, 2nd Ed., 2nd Revision, American Institute of Steel Construction, Chicago, IL.

Ali, H. E. M. and Abdel-Ghaffar, A. M. (1994), "Seismic Energy Dissipation for Cable-Stayed Bridges Using Passive Devices," *Earthquake Engineering & Structural Dynamics*, Vol. 23, No. 8, pp. 877-893.

Alperovich, L. and Zheludev, V. (1998), "Wavelet Transform as a Tool for Detection of Geomagnetic Precursors of Earthquakes," *Physics and Chemistry of the Earth*, Vol. 23, pp. 965-967.

Arnold, W. F. (1984), "Generalized Eigenproblem Algorithms and Software for Algebraic Riccati Equations," *Proceedings of the IEEE*, Vol. 72, No. 12, pp. 1746-1754.

Bakshi, B. R. and Stephanopoulos, G. (1993), "Wave-Net: a Multiresolution, Hierarchical Neural Network with Localized Learning," *AIChE Journal*, Vol. 39, No.1, pp. 57-81.

Balendra, T., Wang, C. M., and Cheong, H. F. (1995), "Effectiveness of Tuned Liquid Column Dampers for Vibration Control of Towers," *Engineering Structures*, Vol. 17, No. 9, pp. 668-675.

Basu, B. and Gupta, V. K. (1997), "Non-stationary Seismic Response of MDOF Systems by Wavelet Transform," *Earthquake Engineering and Structural Dynamics*, Vol. 26, pp. 1243-1258.

Blevins, R. D. (1984), *Applied Fluid Dynamics Handbook*, Van Nostrand Reinhold Co., New York, NY.

Bossens, F. and Preumont, A. (2001), "Active Tendon Control of Cable-Stayed Bridges: A Large-Scale Demonstration," *Earthquake Engineering & Structural Dynamics*, Vol. 30, No. 7, pp. 961-979.

Burdisso, R. A., Surarez, L. E. and Fuller, C. R. (1994), "Feasibility Study of Adaptive Control of Structures Under Seismic Excitation," *Journal of Engineering Mechanics*, ASCE, Vol. 120, No. 3, pp. 580-592.

Burrus, C. S., Gopinath, R. A., and Guo. H. (1998), *Introduction to Wavelets and Wavelet Transforms: a Primer*, Prentice Hall, Upper Saddle River, NJ.

Chen, H. M., Tsai, K. H., Qi, G. Z., Yang, J. C. S. and Amini, F. (1995), "Neural Network for Structure Control," *Journal of Computing in Civil Engineering*, Vol. 9, No. 2, pp. 168-176.

Christenson, R. E., Spencer, B. F. Jr., Hori, N. and Seto, K. (2003), "Coupled Building Control Using Acceleration Feedback," *Computer-Aided Civil and Infrastructure Engineering*, Vol. 18, No. 1, pp. 3-17.

Chui, C. K. (1992), *An Introduction to Wavelets*, Academic Press, Inc., San Diego, CA.

Chung, L. L., Lin, R. C., Soong, T. T., and Reinhorn, A. M. (1989), "Experiments on Active Control for MDOF Seismic Structures," *Journal of Engineering Mechanics*, ASCE, Vol. 115, No. 8, pp. 1609-1627.

Connor, J. J. (2003), *Introduction to Structural Motion Control*, Prentice Hall, Upper Saddle River, NJ.

Craig, R. R. (1981), *Structural Dynamics: An Introduction to Computer Methods*, Wiley, New York.

Daubechies, I. (1988), "Orthonormal Bases of Compactly Supported Wavelets," *Communications on Pure and Applied Mathematics*, Vol. 41, pp. 909-996.

Daubechies, I. (1992), *Ten Lectures on Wavelets*, Society for Industrial and Applied Mathematics, Philadelphia, PA.

Dorato, P., Abdallah, C., and Cerone, V. (1995), *Linear-Quadratic Control: An Introduction*, Prentice Hall, Englewood Cliffs, NJ.

Dyke, S. J., Spencer, B. F. Jr., Quast, P., Kaspari, D. C. Jr. and Sain, M. K. (1996a), "Implementation of an Active Mass Driver using Acceleration Feedback," *Microcomputers in Civil Engineering*, Vol. 11, No. 5, pp. 305-323.

Dyke, S. J., Spencer, B. F. Jr., Sain, M. K., and Carlson, J. D. (1996b), "Modeling and Control of Magnetorheological Dampers for Seismic Response Reduction," *Smart Materials and Structures*, Vol. 5, No. 5, pp. 565-575.

Dyke, S. J., Turan, G., Caicedo, J. M., Bergman, L. A., and Hague, S. (2000), "Benchmark Control Problem for Seismic Response of Cable-Stayed Bridges," *Proceedings of the Second European Conference on Structural Control*.

Feng, Q. and Shinozuka, M. (1993), "Control of Seismic Response of Bridge Structures using Variable Dampers," *Journal of Intelligent Material Systems and Structures*, Vol. 4, No. 1, pp. 117-122.

Fortner, B. (2001), "Water Tanks Damp Motion in Vancouver High-Rise," *Civil Engineering*, June, p. 18.

Gabor, D. (1946), "Theory of Communication," *Journal of Institute of Electrical Engineers*. London, 93(3), 429-457.

Ghaboussi, J. and Joghataie, A. (1995), "Active Control of Structures using Neural Networks," *Journal of Engineering Mechanics*, ASCE, Vol. 121, No. 4, pp. 555-567.

Ghosh-Dastidar, S. and Adeli, H. (2003), "Wavelet-Clustering-Neural Network Model for Freeway Incident Detection," *Computer-Aided Civil and Infrastructure Engineering*, Vol. 18, No. 5, pp. 325-338.

Ghosh-Dastidar, S. and Adeli, H. (2006), "Neural Network-Wavelet Micro-Simulation Model for Delay and Queue Length Estimation at Freeway Work Zones," *Journal of Transportation Engineering*, ASCE, Vol. 132, No. 2, pp. 331-341.

Ghosh-Dastidar, S., Adeli, H., Dadmehr, N. (2007), "Mixed-band Wavelet-Chaos-Neural Network Methodology for Epilepsy and Epileptic Seizure Detection," *IEEE Transactions on Biomedical Engineering*, Vol. 54, No. 9, September, pp. 1545-1551.

Goupillaud, P., Grossmann, A., and Morlet, J. (1984), "Cycle-octave and Related Transforms in Seismic Signal Analysis," *Geoexploration*, Vol. 23, pp. 85-102.

Grossman, A. and Morlet, J. (1984), "Decomposition of Hardy Functions into Square Integrable Wavelets of Constant Shape," *SIAM Journal on Mathematical Analysis*, Vol. 15, pp. 723-736.

Hanson, R. D. and Soong, T. T. (2001), *Seismic Design with Supplemental Energy Dissipation Devices*, Earthquake Engineering Research Institute (EERI), Oakland, CA.

He, W. L., Agrawal, A. K. and Mahmoud, K. (2001), "Control of Seismically Excited Cable-Stayed Bridge Using Resetting Semiactive Stiffness," *Journal of Bridge Engineering*, Vol. 6, No. 6, pp. 376-384.

Housner, G. W. (1970), "Strong Ground Motion," in *Earthquake Engineering,* Wiegel, R. L. Ed., Prentice-Hall, Inc., Englewood Cliffs, N. J., pp. 75-91.

Housner, G. W., Bergman, L. A., Caughey, T. K., Chassiakos, A. G., Claus, R. O., Masri, S. F., Skelton, R. E., Soong, T. T., Spencer, B. F., and Yao,

J. T. P. (1997), "Structural Control: Past, Present, and Future," *Journal of Engineering Mechanics*, ASCE, Vol. 123, No. 9, Special Issue, pp. 897-971.

Hrovat, D., Barak, P., and Rabins, M. (1983), "Semi-Active Versus Passive or Active Tuned Mass Dampers for Structural Control," *Journal of Engineering Mechanics*, ASCE, Vol. 109, No. 3, pp. 691-705.

Hung, S. L. and Adeli, H. (1991a), "A Model of Perceptron Learning with a Hidden Layer for Engineering Design," *Neurocomputing*, Vol. 3, No. 1, pp. 3-14.

Hung, S. L. and Adeli, H. (1991b), "A Hybrid Learning Algorithm for Distributed Memory Multicomputers," *Heuristics*, Vol. 4, No. 4, pp. 58-68.

Hung, S. L. and Adeli, H. (1993), "Parallel Backpropagation Learning Algorithms on Cray Y-MP8/864 Supercomputer," *Neurocomputing*, Vol. 5, No. 6, pp. 287-302.

Hung, S. L. and Adeli, H. (1994), "A Parallel Genetic/Neural Network Learning Algorithm for MIMD Shared Memory Machines," *IEEE Transactions on Neural Networks*, Vol. 5, No. 6, pp. 900-909.

IBC (2000), *International Building Code 2000*, International Code Council, Falls Church, Virginia.

Iyama, J. and Kuwamura, H. (1999), "Application of Wavelets to Analysis and Simulation of Earthquake Motions," *Earthquake Engineering and Structural Dynamics*, Vol. 28, pp. 255-272.

Jiang, X. and Adeli, H. (2003), "Fuzzy Clustering Approach for Accurate Embedding Dimension Identification in Chaotic Time Series," *Integrated Computer-Aided Engineering*, Vol. 10, No. 3, pp. 287-302.

Jiang, X. and Adeli, H. (2004), "Wavelet Packet-Autocorrelation Function Method for Traffic Flow Pattern Analysis," *Computer-Aided Civil and Infrastructure Engineering*, Vol. 19, No. 5, pp. 324-337.

Jiang, X. and Adeli, H. (2005a), "Dynamic Wavelet Neural Network for Nonlinear Identification of Highrise Buildings," *Computer-Aided Civil and Infrastructure Engineering*, Vol. 20, No. 5, pp. 316-330.

Jiang, X. and Adeli, H. (2005b), "Dynamic Wavelet Neural Network Model for Traffic Flow Forecasting," *Journal of Transportation Engineering*, ASCE, Vol. 131, No. 10, pp. 771-779.

Jiang, X. and Adeli, H. (2007), "Pseudospectra, MUSIC, and Dynamic Wavelet Neural Network for Damage Detection of Highrise Buildings," *International Journal for Numerical Methods in Engineering*, Vol. 71, No. 5, July, pp. 606-629.

Jiang, X. and Adeli, H. (2008a), "Dynamic Fuzzy Wavelet Neuroemulator for Nonlinear Control of Irregular Highrise Building Structures, *International Journal for Numerical Methods in Engineering*, Vol. 74, No. 7, May, pp. 1045-1066.

Jiang, X. and Adeli, H. (2008b), Neuro-Genetic Algorithm for Nonlinear Active Control of Highrise Buildings, *International Journal for Numerical Methods in Engineering*, Vol. 75, No. 8, pp. 770-786.

Kareem, A. (1994), "The Next Generation of Tuned Liquid Dampers," *Proceedings of First World Conference on Structural Control*, Los Angeles, California, USA, pp. FP5-19-FP5-28.

Kareem, A. and Kline, S. (1995), "Performance of Multiple Mass Dampers under Random Loading," *Journal of Structural Engineering*, Vol. 121, No. 2, pp. 348-361.

Karim, A. and Adeli, H. (2002a), "Comparison of the Fuzzy-Wavelet RBFNN Freeway Incident Detection Model with the California

Algorithm," *Journal of Transportation Engineering*, ASCE, Vol. 128, No. 1, pp. 21-30.

Karim, A. and Adeli, H. (2002b), "Incident Detection Algorithm using Wavelet Energy Representation of Traffic Patterns," *Journal of Transportation Engineering*, ASCE, Vol. 128, No. 3, pp. 232-242.

Karim, A. and Adeli, H. (2003), "Fast Automatic Incident Detection on Urban and Rural Freeways using the Wavelet Energy Algorithm," *Journal of Transportation Engineering*, ASCE, Vol. 129, No. 1, pp. 57-68.

Kim, H. and Adeli, H. (2004), "Hybrid Feedback-LMS Algorithm for Structural Control," *Journal of Structural Engineering*, ASCE, Vol. 130, No. 1, pp. 120-127.

Kim, H. and Adeli, H. (2005a), "Wind-Induced Motion Control of 76-Story Benchmark Building using the Hybrid Damper-TLCD System," *Journal of Structural Engineering*, ASCE, Vol. 131, No. 12, pp. 1794-1802.

Kim, H. and Adeli, H. (2005b), "Hybrid Control of Smart Structures using a Novel Wavelet-based Algorithm," *Computer-Aided Civil and Infrastructure Engineering*, Vol. 20, No. 1, pp. 7-22.

Kim, H. and Adeli, H. (2005c), "Wavelet-Hybrid Feedback LMS Algorithm for Robust Control of Cable-Stayed Bridges," *Journal of Bridge Engineering*, ASCE, Vol. 10, No. 2, pp. 116-123.

Kim, H. and Adeli, H. (2005d), "Hybrid Control of Irregular Steel Highrise Building Structures under Seismic Excitations," *International Journal for Numerical Methods in Engineering*, Vol. 63, No. 12, pp. 1757-1774.

Kurata, N., Kobori, T., Takahashi, M., Niwa, N. and Midorikawa, M. (1999), "Actual Seismic Response Controlled Building with Semi-Active Damper System," *Earthquake Engineering & Structural Engineering*, Vol. 28, No. 11, pp. 1427-1447.

Lee-Glauser, G. J., Ahmadi, G., and Horta, L. G. (1997), "Integrated Passive/Active Vibration Absorber for Multistory Buildings," *Journal of Structural Engineering*, ASCE, Vol. 123, No. 4, pp. 499-504.

Lewalle, J. (1995), http://www.ecs.syr.edu/faculty/lewalle/tutor/tutor.html

Lewis, F. L. and Syrmos, V. L. (1995), *Optimal Control*, Wiley, New York, NY.

Liang, S., Li, Q., and Qu, W. (2000), "Control of 3-D Coupled Responses of Wind-Excited Tall Buildings by a Spatially Placed TLCD System," *Wind & Structures*, Vol. 3, No. 3, pp. 193-207.

Lyubushin, A. (1999), "Wavelet-aggregated Signal in Earthquake Prediction," *Earthquake Research in China,* 13(1), pp. 33-43.

Mallat, S. G. (1989), "A Theory for Multiresolution Signal Decomposition: The Wavelet Representation," *IEEE Transactions on Pattern Analysis and Machine Intelligence*, Vol. 11, No. 7, pp. 674-693.

Matlab (1999), *SIMULINK: Dynamic System Simulation for Matlab*, Mathworks Inc, Natick, MA.

Matlab (2000), *Control System Toolbox for Use with Matlab*, Mathworks Inc, Natick, MA.

Meyer, Y. (1993), *Wavelets: Algorithms & Applications*, Society for Industrial and Applied Mathematics, Philadelphia, PA.

Miyamoto, H. K. and Scholl, R. E. (1998), "Steel Pyramid," *Modern Steel Construction*, November, pp. 42-28.

Mokha, A. S., Amin, N., Constantinou, M. C., and Zayas, V. (1996), "Seismic Isolation Retrofit of Large Historic Building," *Journal of Structural Engineering*, Vol. 122, No. 3, pp. 298-308.

Newland, D. E. (1993), *An Introduction to Random Vibrations, Spectral and Wavelet Analysis*, Wiley, New York, NY.

Oonincx, P. J. (1999), "A Wavelet Method for Detecting S-waves in Seismic Data," *Computational Geosciences*, Vol. 3, pp. 111-134.

Prakah-Asante, K. O. and Craig, K. C. (1994), "The Application of Multi-Channel Design Methods for Vibration Control of an Active Structure," *Smart Material and Structures*, Vol. 3, No. 3, pp. 329-343.

Rao, R. M. and Bopardikar, A. S. (1998), *Wavelet Transforms: Introduction to Theory and Applications*, Addison-Wesley, Reading, MA.

Sadek, F. and Mohraz, B. (1998), "Semiactive Control Algorithms for Structures with Variable Dampers," *Journal of Engineering Mechanics*, ASCE, Vol. 124, No. 9, pp. 981-990.

Sadek, F., Mohraz, B. and Lew, H. S. (1998), "Single- and Multiple-Tuned Liquid Column Dampers for Seismic Applications," *Earthquake Engineering & Structural Dynamics*, Vol. 27, No. 5, pp. 439-463.

Sakai, F., Takaeda, S., and Tamaki, T. (1989), "Tuned Liquid Column Damper – New Type Device for Suppression of Building Vibrations," *Proceedings of International Conference on Highrise Buildings*, Nanjing, China, pp. 926-931.

Sakai, F., Takaeda, S., and Tamaki, T. (1991), "Tuned Liquid Column Damper (TLCD) for Cable-Stayed Bridges," *Proceedings of Specialty Conf. Invitation in Cable-Stayed Bridges*, Fukuoka, Japan, pp. 197-205.

Saleh A. and Adeli, H. (1994), "Parallel Algorithms for Integrated Structural and Control Optimization," *Journal of Aerospace Engineering*, ASCE, Vol. 7, No. 3, pp. 297-314.

Saleh, A. and Adeli, H. (1997), "Robust Parallel Algorithms for Solution of the Riccati Equation," *Journal of Aerospace Engineering*, ASCE, Vol. 10, No. 3, pp. 126-133.

Saleh, A. and Adeli, H. (1998a), "Optimal Control of Adaptive Building Structures under Blast Loading, *Mechatronics*, Vol. 8, No. 8, pp. 821-844.

Saleh, A. and Adeli, H. (1998b), "Optimal Control of Adaptive/Smart Multistory Building Structures," *Computer-Aided Civil and Infrastructure Engineering*, Vol. 13, No. 6, pp. 389-403.

Samali, B., Kwok, K., and Gao, H. (1998), "Wind Induced Motion Control of a 76 Story Building By Liquid Dampers" *Proceedings Of Second World Conference on Structural Control*, Vol. 2, pp. 1473-1480, John Wiley & Sons, New York, NY.

Samant, A. and Adeli, H. (2000), "Feature Extraction for Traffic Incident Detection using Wavelet Transform and Linear Discriminant Analysis," *Computer-Aided Civil and Infrastructure Engineering*, Vol. 15, No. 4, pp. 241-250.

Samant, A. and Adeli, H. (2001), "Enhancing Neural Network Incident Detection Algorithms using Wavelets," *Computer-Aided Civil and Infrastructure Engineering*, Vol. 16, No. 4, pp. 239-245.

Schemmann, A. G. and Smith, H. A. (1998a), "Vibration Control of Cable-Stayed Bridges, Part 1: Modeling Issues," *Earthquake Engineering & Structural Dynamics*, Vol. 27, No. 8, pp. 811-824.

Schemmann, A. G. and Smith, H. A. (1998b), "Vibration Control of Cable-Stayed Bridges, Part 2: Control Analyses," *Earthquake Engineering & Structural Dynamics*, Vol. 27, No. 8, pp. 825-843.

Simiu, E. and Scanlan, R. H. (1996), *Wind Effects on Structures: Fundamentals and Applications to Design*, John Wiley, New York, NY.

Singh, M. P. and Matheu, E. E. (1997), "Active and Semi-active Control of Structures under Seismic Excitation," *Earthquake Engineering & Structural Dynamics*, Vol. 26, No. 2, pp. 193-213.

208

Sirca, G. and Adeli, H. (2004), "A Neural Network-Wavelet Model for Generating Artificial Accelerograms," *International Journal of Wavelets, Multiresolution, and Information Processing*, Vol. 2, No. 3, pp. 217-235.

Skelton, R. E. (1988), *Dynamic Systems Control: Linear Systems Analysis and Synthesis*, Wiley, New York, NY.

Soong, T. T. (1990), *Active Structural Control: Theory and Practice*, Wiley, New York, NY.

Soong, T. T. and Constantinou, C., Eds. (1994), *Passive and Active Structural Vibration Control in Civil Engineering*, Springer-Verlag, New York, NY.

Soong, T. T. and Reinhorn, A. M. (1993), "An Overview of Active and Hybrid Structural Control Research in the U.S.," *Structural Design of Tall Buildings*, Vol. 2, No. 3, pp. 192-209.

Spencer, B. F. Jr., Dyke, S. J. and Deoskar, H. S. (1998), "Benchmark Problems in Structural Control: Part I — Active Mass Driver System," *Earthquake Engineering & Structural Engineering*, Vol. 27, No. 11, pp. 1127-1139.

Spencer, B. F. Jr., Suhardjo, J., and Sain, M. K. (1994), "Frequency Domain Optimal Control Strategies for Aseismic Protection," *Journal of Engineering Mechanics*, ASCE, Vol. 120, No. 1, pp. 135-158.

Stein, G. and Athans, M. (1987), "The LQG/LTR Procedure for Multivariable Feedback Control Design," *IEEE Transactions on Automatic Control*, Vol. AC32, No. 2, pp. 105-114.

Suhardjo, J., Spencer, B. F. Jr., and Kareem, A. (1992), "Frequency Domain Optimal Control of Wind-Excited Buildings," *Journal of Engineering Mechanics*, ASCE, Vol. 118, No. 12, pp. 2463-2481.

Symans, M. D. and Constantinou, C. (1997), "Seismic Testing of a Building Structure with a Semi-Active Fluid Damper Control System," *Earthquake Engineering & Structural Dynamics*, Vol. 26, No. 7, pp. 759-777.

Symans, M. D. and Constantinou, C. (1999), "Semi-active Control Systems for Seismic Protection of Structures: A State-of-the-art Review," *Engineering Structures*, Vol. 21, No. 6, pp. 469-487.

Tabatabai, B. and Mehrabi, A. B. (2000), "Design of Mechanical Viscous Dampers for Stay Cables," *Journal of Bridge Engineering*, Vol. 5, No. 2, pp. 114-123.

Tarrab, M. and Feuer, A. (1988), "Convergence and Performance Analysis of the Normalized LMS Algorithm with Uncorrelated Gaussian Data," *IEEE Transactions on Information Theory*, Vol. 34, No. 4, pp. 680-691.

Teramura, A. and Yoshida, O. (1996), "Development of Vibration Control System using U-shaped Water Tank," *Proceedings of the Eleventh World Conference on Earthquake Engineering*, Pergamon, Paper No. 1343.

Turan, G., Voulgaris, P., and Bergman, L. (2002), "μ-Synthesis Control of a Cable-Stayed Bridge Against Earthquake Excitation," *Proceedings of the Third World Conference on Structural Control*, Como, Italy. Also available at http://wusceel.cive.wustl.edu/quake

Villaverde, R. and Marin, S. C. (1995), "Passive Seismic Control of Cable-Stayed Bridges with Damped Resonant Appendages," *Earthquake Engineering & Structural Dynamics*, Vol. 24, No. 2, pp. 233-246.

Wang, M. L. and Wu, F. (1995), "Structural System Identification using Least Mean Square (LMS) Adaptive Technique," *Soil Dynamics and Earthquake Engineering*, Vol. 14, No. 6, pp. 409-418.

Warnitchai, P., Fujino, Y., Pacheco, B. M. and Agret, R. (1993), "Experimental Study on Active Tendon Control of Cable-Stayed Bridges," *Earthquake Engineering & Structural Dynamics*, Vol. 22, No. 2, pp. 93-111.

Widrow, B. J. and Stearns, S. D. (1985), *Adaptive Signal Processing*, Prentice-Hall, Englewood Cliffs, NJ.

Won, A. Y. J., Pires, J. A., and Haroun, M. A. (1996), "Stochastic Seismic Performance Evaluation of Tuned Liquid Column Dampers," *Earthquake Engineering & Structural Dynamics*, Vol. 25, No. 11, pp.1259-1274.

Wu, M. and Adeli, H. (2001), "Wavelet-Neural Network Model for Automatic Traffic Incident Detection," *Mathematical and Computational Applications*, Vol. 6, No. 2, pp. 85-96.

Yalla, S. K., and Kareem, A. (2000), "Optimum Absorber Parameters for Tuned Liquid Column Dampers," *Journal of Structural Engineering*, Vol. 126, No. 8, pp. 906-915.

Yalla, S. K., Kareem, A., and Kantor, J. C. (2001), "Semi-active Tuned Liquid Column Dampers for Vibration Control of Structures," *Engineering Structures*, Vol. 23, No. 11, pp. 1469-1479.

Yang, G. and Satoh Y. (2001), *Java-Powered Simulator for Structural Vibration Control*, http://www.nd.edu/~quake/java/sin.html.

Yang, G., Jung, H. J. and Spencer, B. F. Jr. (2001), "Dynamic Modeling of Full-Scale MR Dampers for Civil Engineering Applications," *US-Japan Workshop on Smart Structures for Improved Seismic Performance in Urban Region*, Seattle, WA, pp. 14-16.

Yang, J. N. (1982), "Control of Tall Building under Earthquake Excitation," *Journal of Engineering Mechanics Division*, ASCE, Vol. 108, No. EM5, pp. 833-849.

Yang, J. N. and Li, Z. (1991), "Instantaneous Optimal Control of Linear and Nonlinear Structures – Stable Controller," *Technical Report, NCEER-TR-91-0026*, National Center for Earthquake Engineering Research, Buffalo, NY.

Yang, J. N., Agrawal, A. K., Samali, B., and Wu, J. C. (2000), "A Benchmark Problem for Response Control of Wind-Excited Tall Buildings," *Proceedings of Fourteenth Engineering Mechanics Conference*, ASCE, Austin, Texas, Also available at: http://www-ce.engr.ccny.cuny.edu/people/faculty/agrawal/benchmark.html.

Yang, J. N., Akbrapour, A. and Ghaemmaghami, P. (1987), "New Optimal Control Algorithms for Structural Control," *Journal of Engineering Mechanics*, ASCE, Vol. 113, No. 9, pp. 1369-1386.

Yang, J. N., Wu, J. C., Samali, B., and Agrawal, A. K. (1998), "A Benchmark Problem for Response Control of Wind-Excited Tall Buildings" *Proceedings Of Second World Conference on Structural Control*, Vol. 2, pp. 1407-1416, John Wiley & Sons, New York, NY.

Yao, J. T. P. (1972), "Concept of Structural Control," *Journal of Structural Division*, ASCE, Vol. 98, No. ST7, pp. 1567-1574.

Youssef, N., Nuttall, B., Hata, O., Tahtakran, O., and Hart, G. C. (2000), "Los Angeles City Hall," *Structural Design of Tall Buildings*, Vol. 9, No. 1, pp. 3-24.

Zhang, R.H. and Soong, T. T. (1992), "Seismic Design of Viscoelastic Dampers for Structural Applications," *Journal of Structural Engineering*, ASCE, Vol. 118, No. 5, pp. 1375-1392.

Zhou, Z. and Adeli, H. (2003a), "Time-frequency Signal Analysis of Earthquake Records using Mexican Hat Wavelets," *Computer-Aided Civil and Infrastructure Engineering*, Vol. 18, No. 5, pp. 379-389.

Zhou, Z. and Adeli, H. (2003b), "Wavelet Energy Spectrum for Time-Frequency Localization of Earthquake Energy," *International Journal of Imaging Systems and Technology*, Vol. 13, No. 2, pp. 133-140.

INDEX